12—

THE CASE AGAINST
HILLARY
CLINTON

THE CASE AGAINST
HILLARY CLINTON

PEGGY NOONAN

ReganBooks
An Imprint of HarperCollins*Publishers*

HarperCollins books may be purchased for educational, business, or sales promotional use. For information please write: Special Markets Department, HarperCollins Publishers Inc., 10 East 53rd Street, New York, NY 10022.

FIRST EDITION

Designed by Philip Mazzone

Printed on acid-free paper

Library of Congress Cataloging-in-Publication Data has been applied for.

ISBN 0-06-039340-8

00 01 02 03 04 ❖/RRD 10 9 8 7 6 5 4 3 2 1

ACKNOWLEDGMENTS

ALL BOOKS ARE CALLED INTO BEING BY LITTLE platoons of friends, professional and otherwise. Sincere thanks to Jim Pinkerton, Marie Brenner, Misty Church, Ransom Wilson, Stephani Young, Bill Sykes, Joni Evans, Sim Johnston, Lisa Johnston, Slade Gorton, Jennifer Sherwood, Cal Thomas, Judith Regan, Susan and Vincent Woodbury, Amy Woodbury, Ernest Pomerantz, Genevieve Koppel, Will Rahn, Anne Reingold, Lisa Schwarzbaum, Max Boot, Eric Zohn, Jim Fox, the Levinson family, and the National Data Conversion Institute—Michael, Luchino, Steven and Teresa.

TO ELEANOR ROOSEVELT

THE CASE AGAINST
HILLARY CLINTON

PREFACE

IT WAS A ROAR.

It wasn't just cheering and applause, it was a roar. And the lights and the laughter and the chants—

Up front against the rope below the stage in the huge ballroom of the Hilton they were packed so close it looked like a conga line, and when she saw it she threw back her head and laughed.

She had just been introduced. She was standing at the podium smiling and waving, her blond hair shining in the lights. She'd stuck to the uniform, smart girl, only tonight the pantsuit was blue-black silk draped delicately over satin pumps, and the blouse, silk also, was soft blue, like a robin's egg.

She smiles—crinkly eyed, sparkling. She looks attractive and maternal, like a nice woman who's a strong woman. She's never looked so beautiful. She's never *been* beautiful, not till this year. But now all is changed.

She gives a little nod, touches the microphone lightly, looks out at the mob, and says,

"Thank you. Thank you so much."

But you couldn't hear the last words because they erupted again at the sound of her voice.

"Thank you."

The heady din.

She laughed again, and I knew what was happening. I have seen this moment before, the moment when the victor says thank you and the crowd roars, and says thank you again to more roar.

This is the moment when politicians in a close race get to do something they haven't been able to do in months, something they're hungry to do but have been too ragged around to do, too busy reacting, chatting, strategizing, laughing, speaking, extroverting, moving, charming, and hugging, to do. And that is: think.

Just think. Without someone handing you a note or a phone or a poll or a baby.

I knew what she was thinking as she surveyed the crowd, smiling. She was thinking, *I didn't know Al Sharpton could jump.*

And the thought made her laugh. He was over there with Harold, jumping with joy. He'd been a hard one, but by the end they'd gotten him aboard.

You've probably noticed what politicians do when a crowd gets like this. They smile shyly, and find ways to keep the roar going to show how much they're loved.

She sees, peripherally, that over in the corner where the tripods and cameras and TV reporters are, something has changed. The red lights are on—the reporters are talking into the cameras now, they've got one hand on their ears to keep the earpieces in so they can do Q&A with Chuck or Sue, but right now they're reporting live from the scene, "It's just jubilant here, Chuck, as you can see. Hillary Clinton is about to speak but the crowd won't let her. It was a close race, it was only apparently moments ago that the mayor conceded, and Mrs. Clinton is about to make her victory statement—if the crowd will let her. So far they're intent on showing her how happy they are. Let's just let the camera pan here a little and show the scene, about 3,000 people in a crammed ballroom in the Hilton ..."

She knows what they're saying—this is not, as she will inform you, her first time at the rodeo—and she strings it out by shaking her head, stretching her right arm and pointing her index finger at friends in the back of the crowd.

She learned it from her husband: the Triumphant Pointing Pose. You stand on a stage surrounded by hordes and you point outward and it looks commanding. It's a good picture for *Time* and *Newsweek*. And let me tell you, I've whipped my head around to follow the president's finger when he does it and believe me: There's no one there. He's pointing at the air. He just pretends someone's there.

She points at the air, does a thumbs-up, takes a quick furtive look at her notes.

And thinks: *Remember to thank Judith Hope. "And Judith, for the best state apparatus in the"—no, change apparatus to people. Remember Charlie Rangel—make it personal. "One day Charlie Rangel came to me and said 'Hillary, I've got an amaz-*

ing idea' . . . *and eighteen long months later that idea, that dream, has come true . . ."* That'll do it.

She doesn't know it but she nods.

And waves some more.

Now she's in full-bore free association, thinking about what it all means. She is thinking:

This is my night. This time it's me in the lights. Thirty years of the long march from Wellesley, thirty years since I gave the speech that captured the yearnings of a generation. Poor Ed Brooke never knew what hit him— I stood up in my graduation gown and took on a U.S. Senator. Now I am a U.S. Senator . . .

A quarter century since I met Bill, and the odyssey began . . .

And now I've won, and I've beaten more than Rudy. I've beaten Bill.

He got it and blew it. Now I'll show him how it's done.

"Thank you. Thank you so much."

More roar. They're doing their part. It's only campaign Kabuki, and yet no one—the people, the press, the pols—no one ever tires of it, this enactment of joy at the victory and love for the victor.

"I thank you so much."

What's that they're chanting—"Six more years! Six more years!"

She laughs, and shakes her head.

Actually I might be moving on in four, but we'll see how that goes. With Bush's victory tonight I'm the most famous Democrat in the country, number one on line for 2004. But who knows, life is long. I'm only fifty-three. Reagan was almost twenty years older than that when he walked into the White House for the first time . . .

She looks up at the seats in the balcony. For all candidates of a certain age, men and women, this is the beauty shot, the one where you look up and the layer of subcutaneous fat below your chin disappears, the soft jawline gets firm and you look . . . young again, and glistening.

She waves and thinks, *I'm home now.*

They say I don't have a home, rootless, bouncing—Illinois, Massachusetts, Connecticut, Washington, Arkansas, Washington, never settling in.

Well, this is my home. In the lights, with my power.

A whole future opening up.

I have survived all.

That's why—is anyone noticing?—I picked the music. They think it's just happy-campaign-disco-boom they're hearing but it's Gloria Gaynor, "I Will Survive." "Dya think I'd crumble, dya think I'd lay down and die? Oh no not I—I will survive . . ." Susan Thomases told me no one would hear it and no one would get it. Well, I get it. My message to me.

All right, three and a half minutes—it's gone on enough, we'll lose the networks, they'll go back to Austin.

"Thank you. Thank you."

She puts her right hand up. Still smiling, but there's something in her face. It's time to stop now.

"There are so many people to thank. My wonderful family . . . "

A new burst of applause.

Bill will be watching, at home, upstairs. We arranged it months ago, the post-millennium seminars on welfare in the twenty-first century. The last few weeks of the campaign he started scaring people; they started thinking he was running. And every time he gets on stage with me he can't help himself, he hogs the spotlight, talks too long, and leaves me standing there with the adoring gaze about to crack off my face. And if I stop smiling and start to look distracted the press makes a story of it—"As the president spoke the first lady looked on with boredom—Divorce at eleven." Then I finally get to speak and he stands there with his adoring look, only he loves it—he stands there with perfect posture and nods and works the muscles in his face, and I'm talking HMOs and I look up to make a point and everyone in the audience is looking at . . . glamour man. And he knows it. And knows I know it. "They love me." Well, tonight they love someone else . . .

"I want to thank my wonderful daughter, Chelsea, who's here with me tonight. I told her maybe just this once she could skip school. And I thank my husband, the president, who is in Washington tonight, doing the work of the people . . ."

Cheers.

Small family, getting smaller. Mom's at home, probably listening to right-wing radio. My brother's in Russia doing business. My old friends . . . can't have them here now, too much baggage. Harry is Travelgate. Web Hubbell is—well, Webgate. The old Arkansas gang pretty much gone. Vince Foster gone . . .

"But most of all I thank you, for joining my crusade, for giving me your hearts and your hands, and I promise you, I will never forget it."

She talked about Eleanor Roosevelt then, and she talked about what she would stand for in the Senate—health care for all, schools that ready our children for the twenty-first century . . .

It wasn't bad, but we'd heard it before, and reporters and even some of the crowd started looking around. I fastened on something I'd heard about and hadn't believed until I saw it there on a monitor over in the corner. Guest election commentators on Fox, and chattering happily to Tony Snow is a laughing, dark-haired woman named Monica Lewinsky. I looked over at Hillary. Her eyes could see the monitor too. But she betrayed nothing.

"And so—this has been a long battle, a long campaign, and an important moment, I think, in our national life. And I thank all of you for your overwhelming support, and I invite you all to stay as long as you like, and I'm going to try and thank every one of you. So thank you, my friends. And thank you, New York."

She waved triumphantly, and hugged Mandy and James and Harold and Susan. And as she made her way off the stage she was surrounded, engulfed by people.

That's when I saw it.

Hillary's personal aide, Capricia Marshall, runs up to her with a portable phone, the little antenna pulled out and the flap down. She mouths, "The president."

Hillary is hugging people, and looks at Capricia.

"The president—he says to hook the phone up to the public address system so everyone can hear him congratulate you and New York, he says it'll make a nice moment."

Hillary smiles and says "Give," and takes a hug from Nita Lowey, and pushes a button on the phone and gives it back to Capricia, who took it, confused. She looks.

Hillary had pressed End instead of Send.

And she left Capricia standing there with a dead phone in her hand as she waded into the crowd, walking into it with energy and love, as if she were entering a new and not frightening future.

I think that victory night, that celebration of Hillary Clinton's winning Senate run, may well happen, all of it, beginning to end.

I do not think it should. I hope it doesn't. But one of the reasons I think it might—just one—is that Hillary Clinton really is, as they say, one of those striking modern media figures to whom people seem to . . . bring themselves. They bring aspects of their lives, parts of their experience, and they project those aspects and dramas onto her. She comes to symbolize things for people, as if she stands for certain facts in their lives. Part of the reason for this is our therapeutic culture, which encourages such projections; part of it is due to the nature of modern fame—those who are famous now are not just famous, they're the wallpaper, all around us, on every news show, in every face. But part of the reason for Hillary's iconic power is that she and her husband have lived a public life of outsized personal drama marked by mysteries and betrayals and bitterness and accusations. There's something for almost everyone. Women who have been abused and humiliated by a man are said to see her as a fellow survivor; her victory is their redemption. Feminists see her as a woman operating in the world against the odds; her triumph is theirs. (They also see in her their own political assumptions; she will carry their program forward.) Some middle-aged boomers see in her the last rise of the ethos of the sixties, the ethos of their youth; if she succeeds it means their era, and their investment in it, had meaning.

I see things, too. I look at Mrs. Clinton and see the kneesocked girl in the madras headband, the Key Club president who used to walk into the bathroom in Rutherford High School, wrinkle her nose at the tenth-grade losers leanings against the gray tile walls, leave, go down the hall, and mention to a teacher that they're smoking in the girls' room again.

That's my own private Hillary, or at least one aspect of her.

Here's another—a moment that to me quite captures her. It is January 2000, and the First Lady is on *Letterman*. Her much-anticipated appearance is going well; they're laughing and trading quips. Then Letterman pulls out a list of questions about New York. Mrs. Clinton shows a certain anxiety; her face gets the watchful, listening look people get when they know they're about to be challenged, and in public. She agrees to try and answer, but seems slightly reluctant.

Watching at home, I couldn't help it: I felt nervous for her.

Letterman asks the name of the state bird. "The bluebird," she says, to applause. The state tree? Hillary gets a *Jeopardy* Daily Double look. "Um, the maple." But *which* maple? She blinks, thinking hard. "Let's see, there's the red maple, the sugar maple . . ." That's it, says Letterman, to more applause. On it went—how many counties are there in the state of New York? —and she got every question right. It was impressive. It showed the people of New York that she knows more about us than we knew.

And then the next day it turned out she had seen the questions in advance. She'd been given "a sneak peek," according to her spokesman.

That in itself wasn't shocking. What was shocking was the Daily Double look, the straining gamely for answers. It was all acting. *It was all a fake.* And it was so convincing.

It was so . . . Hillary.

For those who are inclined to react to Hillary as an icon, it's good to remember that she is a person—she's had experiences, she has a record, she has a character and a personality and a history. And it is these things on which she is best judged.

This book is a polemic—an argument for a point of view, in this case a perspective highly critical of Mrs. Clinton. It is based on the public record; it argues against her continued rise to positions of public power. It argues, in fact, that the Clintons' eight years are and should be over, that they have earned retirement, merit it, and that they should, as *The Economist* said of Mr. Clinton eighteen months ago, "Just Go Away."

Critics of the Clintons get labeled "Clinton haters." The purpose of calling them haters is to suggest that they are so blinded by animus that they cannot see, so unbalanced by anger that they cannot be fair.

Christopher Hitchens, the author of a stinging indictment of Bill Clinton entitled *No One Left to Lie To*, is one of those at whom this charge has been leveled. His answer is also mine. "I do not hate them," he said, "I have contempt for them, which is different." There are things I like about Mrs. Clinton, her fortitude and resourcefulness being two. I wish her a long life with good health, much friendship, and many grandchildren. But I do not wish to see her succeed in continuing Clintonism in our national life.

I am grateful to her for her recent invitation to all of us. She has been asking to hear the views of New Yorkers as she proceeds on her listening tour. I was born in Brooklyn, grew up on Long Island, live in Manhattan, and have been a New Yorker for most of my life. And so I offer my thoughts, as a citizen of New York. I hope she hears them.

CHAPTER ONE

I

THERE ARE MOMENTS WHEN I AM STILL AMAZED that a first lady of the United States is scaling down her role, in her words, to run for the U.S. Senate. I don't know why I'm surprised—we've all gotten used to the Clintons' dramas, to all the twists and turns on the hair-raising ride. But the idea that she would abandon her position, which is really a job, and an elevated one, and start hopping on government planes to come to my state and tell me what we need . . .

When, a year ago now, it became clear that she was going to run in New York I sat back in wonder, like everyone else.

I thought: For her to say to a state that she had no connection to, no history with, no previously demonstrated interest in—for her to say to a state whose greatest city she has used for seven years as her own personal cash machine, tying up traffic and inconveniencing millions as she trolls, relentlessly, for campaign money—for her to say to this place full of gifted residents that she deserves to be its senator is an act of such mad boomer selfishness and narcissism that even from the Clintons it was a thing of utter and breathtaking gall.

And then there came the moment when I realized: *They're never going to leave.* Other presidents and first ladies do their work, leave their imprint, and make a graceful exit, departing the stage and attempting to become, if they were not already, wise, high-minded, and fair. But the Clintons—they'll stay until the last footlight fizzles and pops, and then we'll have to wrestle them to the floor of the stage.

I called a friend, a great liberal of the city, a Democrat of forty years, and caught her mid-sputter. "To think of all they've put us through—and now they won't even go away. Who are these people, and why do they think they are necessary to us?" Another, a journalist and Democrat, e-mailed from work when I asked her reaction. "This is how I feel: Lady, keep your hands off my state."

And they are Mrs. Clinton's base.

I was wonderfully angry those first few days, in the spring of '99. I asked everyone I bumped into what they thought, and no one assumed Mrs. Clinton's motives were elevated; there were no choruses of "She is concerned about us" or "She wants to help." Conservatives said she was launching her candidacy to fill the vacuum in her life with our money and our freedom; liberals that she needs therapy after years with that brute, and New York makes a good couch; moderates assumed she needs a place to hang her hat while she ponders her next move.

But then I experienced what everyone experiences with a Clinton story. Within forty-eight hours I had absorbed the new reality and was calculating her prospects and imagining her strategy. It was now all . . . just a fact to me. Not an outrage, just a fact.

We have learned to absorb the Clintons and their many shocks; they have taught us to absorb the brazen, to factor it in and in time discount it. And I suspect they are fully aware of this, that they have learned a number of things in their life in politics, but one of the biggest is this: They can do anything. They are used to the tumbling rhythms of public acceptance: the gasps of shock, followed by the edgy discussion on *Hardball* followed by the earnest discussion on *Wolf Blitzer* followed by the enthusiastic discussion on *Geraldo*. The Clintons watch the news wheel turn, grinding down the pebbles in their path: *Let 'em yell, let 'em send their anger into the air, where it dissipates. When they're done talking I'll still be here.*

There is, always, something admirable in such human toughness. But never has the admirable been so fully wedded to the appalling, never in modern American political history has such tenacity and determination been marshaled to achieve such puny purpose: the mere continuance of Them. Would that they had marshaled their resources to help somebody or something else— their country, the poor, the national defense.

II

And of course she may well win. Republicans hope that in the rigors of the campaign her essential nature will emerge and, in the words of Grover Norquist, the soul of an East German border guard will pop out. They think her inner prison matron will escape and start disciplining the people on the rope line—*No pictures here, buddy, can't you see the sign?* But I've seen her work rooms large and small, seen her on the campaign trail, and she will be a pro, articulate, smooth, and smiling. Babies laugh in her arms.

She is a star, and New York likes stars. She will have the passionate support of the unions and interest groups, and New York has unions and interest groups in abundance. She has a human shield in the Secret Service, a twelve-foot cordon of safety, and won't be as exposed as other candidates. She'll raise a lot of money with ease, especially money outside New York, and won't even have to spend it like other candidates because she'll put every cost she can on the public dime. "Mrs. Clinton says she doesn't want to fly on a government plane, but the Secret Service insists." "Mrs. Clinton says she didn't want the pool, but the Secret Service insists." "Mrs. Clinton says she doesn't like the Chanel body moisturizer, but the Secret Service insists."

There are those who say she isn't tough enough, that she's used to being treated like a queen and handled with kid gloves; she won't know how to put a game face on. But a game face is what she's been wearing for years, and she is plenty tough enough. One example speaks for many: In the 1990 Arkansas gubernatorial campaign she humiliated her husband's opponent, Tom MacRae, by barging into a news conference he was holding at the capitol in Little Rock and taking him on for criticizing her husband. "Get off it, Tom . . . Give me a break!" she scolded. Paul Greenberg of the *Arkansas Democrat-Gazette* later wrote, "Miss Hillary seemed to relish sandbagging her husband's opponents," and succeeded with MacRae because "Poor Tom found himself at a disadvantage: He's a gentleman. And he didn't quite know how to respond. But there was little doubt that Miss Hillary had savored the fight."

She claimed that she'd come upon MacRae's news conference by accident. Later it emerged that she'd come up with the idea to stage the confrontation the day before, during a strategy session.

She will have top-flight media. She will have a war room— the Clintons always have a war room, and it was Mrs. Clinton who ordered the creation of the first one, in Little Rock, and named it. It will be staffed by sharp-elbowed spinners making their bones before going on to reporting jobs at *GMA*.

She has a good shot. She will be a serious candidate. She may be the next senator from New York.

And this is not good.

Because this is not a single stray bid for public office. This is an attempt to continue in American history the ethos, style, and character of the Clinton Administration. This is the continuance of Clintonism. Hillary Clinton's candidacy is a product being locally test-marketed for national consumption: It is the beginning of Hillary for president. It is the last big I.P.O. of the Clinton era; invest now and it will pay off in four years or eight, when we return to the White House. This is a candidacy that will decide whether the Clintons should be advanced in public life and, through that advancement, continue to have a strong impact on what happens in American life in the next century.

The Clinton camp has not attempted to hide this. Mrs. Clinton's friends and staff have told journalists from Tim Russert to Bob Novak that she sees the Senate seat as a stepping-stone to the White House. Russert has said her friends told him it was "a first step to the presidency." In the winter of 1999, Mrs. Clinton said that if elected she will serve all six years of her term, which would take her out of the presidential running in 2004. But Bill Clinton made the same promise to the voters of Arkansas when he ran for reelection in 1990. He said he'd serve his entire term, which would take him out of the presidential running in 1992. Then, as 1992 approached, he announced he'd tour Arkansas and ask the people if he should reconsider his promise. This listening tour, he later announced, convinced him that the people wanted him to run for president in 1992. He didn't break his vow; he just bowed, as always, to the people's wishes. One can well imagine Senator Hillary Clinton, in 2003, doing the same.

So New York is the battle that may decide the war. And so this Senate bid has huge implications, not only for New York State but for the nation.

III

But perhaps you view it only from a local perspective. And perhaps you are sympathetic. After all, you might say, Hillary will be only one senator of a hundred, just one voice in a large disharmonious chorus. Why not give her a pass, let her point of view be represented? She's suffered so. But after seven years some of us think the headline on Mrs. Clinton is not that she has suffered, but that she has made so many others suffer.

And is she really the person to replace Pat Moynihan? Pat

He has done this and she has done this.

And in the doing, they have achieved at least one part of their shared dream: They have made an impact on history, and what they stand for will not soon be forgotten.

II

They believe they are justified in using any means to achieve their ends for a simple and uncomplicated reason. It is that they are superior individuals whose gifts and backgrounds entitle them to leadership. (Bill Clinton signaled this most starkly in 1995 when Mrs. Clinton was being criticized for her failed attempt at health care reform. He said that if every American were as good and hardworking and idealistic as Hillary, this would be a better country.) They believe that they are good and those who oppose them are bad. Therefore it follows that what advances the Clintons is good, and what thwarts, slows, or endangers their rise is bad. And the bad should be fought by whatever means possible.

The Clintons' behavior suggests above all a sharp sense of entitlement to power, a sense confirmed even by the testimony of their friends. Describing Bill Clinton, one friend and supporter told me in the summer of 1999, "He's suffered, the world owes him." When I asked what he meant, the friend said, "He came from a difficult family . . ." In Hillary's case the entitlement appears to arise in part from the fact that she is a woman, a member of a sex that has historically been abused. Therefore talented women should now rise. It is one of Hillary's core convictions that she is very talented, an assessment that appears to be correct. I suspect her sense of entitlement is sharpened, in the words of another longtime colleague and witness to her career, by a sense that she feels "she has a special destiny, that she was chosen by God to do big things." The journalist Michael Kelly touched on this in 1994 when he called her "scarily messianic."

III

One of the most pointed and succinct political observations of the Clinton era was made in 1996 by Senator Bob Kerrey, Democrat of Nebraska, who said, wonderingly, of President Clinton, "Clinton's an unusually good liar. Unusually good. Do you realize that?"

He suffered for that remark, was ostracized to a degree by the triumphant Clintons. But he never retracted what he said, because it was true. (I once asked him if he'd said it, and he laughed and nodded his head. Then I asked him if he'd meant it, and he smiled and gave me a look and nodded his head again.)

The Clintons are unusually brazen. They lie in plain sight, with boldness, with utmost confidence in their ability to carry it off. This again has to do with a sense of entitlement: They seem to feel they *deserve* to carry it off. Bill Clinton lies, as William Safire has noted, "when he doesn't have to." Hillary is, as Safire has observed, "a congenital liar."

Even the way she looks is a fiction. For the past seven years, as we all know, it has changed constantly. This did not, as Bill

Clinton biographer David Maraniss has noted, start in the White House. He told Chris Mathews in 1999 that Hillary has always been "her own Rorschach test." You never know who or what she'll be.

I met her in 1991, before the Clinton presidency. She had brown hair to her shoulders and glasses and wore a shawl. She looked up at people when she spoke to them because in those days she wasn't tall. She talked about her daughter, was friendly and down-home, and struck me as maternal. When I saw her six years later, in 1997, she was a different person. She was in a chic hot-pink suit with bright gold jewelry and sleek black heels. She had short bright blond hair, a formidable air, and made a long and excellent speech without notes.

The point is not that she had changed, people do; or that she's always trying to look better, people do that, too. The point is that she changes so much and so often not only in her look but in the image she is trying to project. The constant parade of new looks reflects an endless attempt to craft a new persona, with new views and new ways to market herself as a commodity. Take-Charge Hillary wows Congress the first time she goes up to testify on health care; Sobbing Hillary tells the boys in the White House that she feels very, very alone right now. Tough Hard Hillary hires barroom bouncers and private eyes; Victim Hillary leaks her anguish to the press. Left-wing Hillary sternly supports a Palestinian state; Pragmatic Pol Hillary says Jerusalem is undivided and Israeli. Traditional First Lady Hillary reminds us to read to our children; Religious Hillary poses in white. Behind-the-Scenes Witch Hillary thinks she owns her staff and doesn't have to be nice to the help. Feminist Hillary is a co-president who creates a national health care plan; Anti-feminist Hillary lets her husband's operatives smear the women he has victimized sexually. Hollywood Hillary, gowned and bejeweled, waves to fans at premieres. Yankees Fan Hillary, Jewish Hillary . . .

You could call her multifaceted, but only in the sense that Sybil, or Eve in *The Three Faces of Eve* were multifaceted. And unlike them she is able to control whom she becomes next. Which is why she's more like Lon Chaney in front of a mirror, building himself a new nose that will reflect a new pose, and a change in the storyline.

She will do whatever works and be whoever she has to be to achieve her objectives.

She wants to be historic but seems unable to distinguish between purposeful ambition and personal vanity. Eleanor Roosevelt gambled her reputation and her security to take her stands; thus she became historic. Hillary wants to be *called* historic.

In the same way she now, by leaving the White House to run for the Senate, wants to be called a pioneer. Lewis and Clark didn't want to be called pioneers; they wanted to find the Northwest Passage. Hillary just wants the label, for the history books.

IV

It is a mistake to underestimate the extent to which Clintonism is based on, informed by, and takes its very shape from the Big Lie.

Clintonism is a way of governance based on a dishonesty so pervasive and all-encompassing, so flamboyantly over the top, that you cannot believe they expect you to believe the bizarre story they just peddled because no one—*no one*—would lie like that or could lie like that.

It is a form of governance that depends on and manipulates the natural fairness and moderation of the American people who, bombarded each day with more Clinton lies, cannot help but absorb, listen, see, hear, and, finally, in a kind of intellectual exhaustion, begin to think . . . *Maybe the Chinese espionage at our nuclear labs really was not his responsibility; maybe Clinton's aides really didn't tell him about it, so he couldn't do anything about it; maybe all the cash he got in '96 from Chinese nationals who are connected to Chinese agents was just a big mistake and had no effect on his decisions. After all, a president wouldn't sell out his country for campaign contributions, would he? Maybe he didn't dodge the draft; maybe he really thinks he came from Hope and not Hot Springs. Maybe he was on the phone with Gennifer Flowers putting down Mario Cuomo because he thought she was a political adviser. Maybe he released the Puerto Rican terrorists because he felt sorry for them and thought they'd been in jail long enough. Maybe he really knew the people who stayed in the Lincoln Bedroom and he didn't just open the White House to wealthy operators who gave him money. Maybe she really didn't know anything about the nine hundred FBI files, or the travel office. Maybe Hillary didn't hear what Suha Arafat was saying; maybe the $100,000 she made in cattle futures was just a lucky stroke; maybe, as Hillary says, ever since Bill began his career he has been "attacked and attacked" because "people sometimes fear change," and "they can see he is a bringer of change," a "change* *agent"* . . . *and the forces of reactionary resistance have silently aligned to keep them from bettering our country.*

Your mind rocks. You make excuses for them time and again because you think, you know from human experience, that no nice educated man in an office like that can be the utterly dishonest person his critics say he is, no nice woman with a kid and a law degree could be the kind of dissembling character her foes insist she is. And then one night when you cannot sleep and your mind begins to wander you think: *Wait, he could be. She could be!* And this thought, this rather frightening, horrific thought, leaves your brain shivering in its skull.

They know people wrestle with who they are, throw up their hands, and in time stop thinking about it because of the sheer impossibility of seeing them for what they are.

The Clintons know that many Americans find it very hard, and maybe impossible, to see them clearly, particularly in a time of such pleasantness and bounty. But, just in case, they remind us again that there's a chicken in every pot and three cars in every carport, not to mention the van.

They know the big lie will work with you much better than the small lie, but they can't resist that, either, a testimony to the sheer force of habit.

You'd think lies, especially the political lies of practiced liars, would be subtle. But theirs are not.

There are many examples. Consider just one: the events surrounding the freeing of the sixteen FALN terrorists whose sentences President Clinton commuted on August 11, 1999.

For ten years starting in 1974 two terrorist groups wanting independence for Puerto Rico—the FALN and another, smaller group—waged a campaign of violence in the United States. They carried out more than a hundred bombings which killed six people and wounded scores of others. In time sixteen defendants were arrested; the indictments brought against them linked them to violent crimes; they were convicted on charges ranging from armed robbery to weapons violations to sedition. All sixteen were believed to have been involved in the terror campaign and all received heavy sentences.

Puerto Rican activists had long pressed for their release, but the terrorists themselves remained unrepentant. The activists began asking President Clinton to release the FALN-ers in 1993, as soon as he took office. He took action on their requests in August 1999.

Why then? Why at all?

He said that the reason for his decision was that the terrorists' sentences were too long under current sentencing guidelines. But the staff director of the U.S. Sentencing Commission soon said that wasn't so; the terrorists could have been tried for treason and received life sentences.

Did the president grant clemency because disinterested professionals in the justice system urged him to? No, they opposed it. The president granted clemency over the unanimous objections of the FBI, whose director, Louis Freeh, warned the president in a memo that the terrorists were unrepentant and likely to return to terrorism if freed; the U.S. Bureau of Prisons; and two U.S. Attorneys. (*Newsweek* later reported that the Bureau of Prisons had audiotapes of some of the terrorists saying that they planned to continue their terror campaign if they were ever freed.)

Was it compassion for the prisoners that motivated the president? Clinton is not known for this particular kind of compassion; he had used the power of presidential pardon only twice in seven years. As governor, he had even allowed a retarded man, Ricky Ray Rector, to be put to death in Arkansas in 1992 rather than face what he imagined would be political heat in the presidential race for being soft on crime. (Ray's execution took place during the Gennifer Flowers scandal, and it has been speculated that the former was meant to take the spotlight off the latter.) There is reason to believe the president associates pardons with politics.

So why did Bill Clinton all of a sudden decide to grant clemency?

The most obvious answer is the most widely held: He was thinking of Hillary's election prospects in New York, which has a high percentage of Puerto Rican voters. Dick Morris, who worked for the Clintons on and off for twenty years, said, "Anyone who doesn't believe the timing, and likely the substance, of Bill's decision was linked to Hillary's courtship of New York's large Puerto Rican vote is too naive for politics."

There is much evidence to support this view, but one small and I suspect not unimportant piece has not previously been addressed in the clemency coverage. It was an appearance on CNBC by Clinton friend and defender Geraldo Rivera in February of 1999, when the first rumors of a Hillary candidacy were beginning to float. Rivera excitedly told a story. He said that he had just talked with the president on the phone and had told him, "'You know, I'd like to invite the first lady to march with me in this summer's Puerto Rican Day parade up Fifth Avenue . . . a million and a half people lining both sides of the streets.'" Bill Clinton, he said, was very interested. "The president instantly said, 'I think she'd like that.' And then, as if to emphasize it, he said, 'I think she'd like that' a second time."

This suggested to Rivera that the rumors of a candidacy were true, and the president was excited about ideas for how to win.

It was six months after this conversation that the president granted clemency to the terrorists. Is it too bold to speculate that that magical number—one and a half million Puerto Rican voters—sparked or sealed his decision?

During the firestorm that followed the clemency announcement, the president insisted that "political considerations played no role" in his decision. He also said that he had not discussed the clemency offer with his wife. The first lady, he said, "didn't know anything about it."

Hillary Clinton also claimed she had "no prior knowledge" of the decision. But only months after the decision, new information regularly emerges that Mrs. Clinton wasn't telling the truth. The *New York Daily News* reported in November 1999 that a leading advocate of clemency sought the first lady's help two months before the clemency was announced. Luis Nieves Falcon, a clemency supporter who, according to the *News*, "attended at least one 1996 White House meeting on the FALN," wrote Mrs. Clinton in June of 1999 saying, "The Puerto Rican community will stand behind your support to this cause. We urge you to support the liberation of 15 Puerto Rican men and women incarcerated in United States federal prisons . . ." The Clinton White House response to the report? Mrs. Clinton gets "thousands of letters each day and obviously doesn't get a chance to see most of the letters herself." Which is not a denial that she got and read the letter, only a statement concerning mail practices. Would her staff hide from her a letter from a political leader who has been to the White House and who is an activist in a major ethnic group in the state she hopes to win?

The Clintons' claim that they never discussed the issue strains credulity in other ways. It is at odds with Mrs. Clinton's repeated assertions over the years that there is nothing they do not discuss. Three months after the clemency decision she was quoted in a long interview in *George* magazine saying, "We talk about everything, and have for as long as we've known each other . . . particularly on issues where we share a common commitment." She said the same in her famous *Talk* magazine interview in September 1999, when she described breakfasting with her husband and discussing education reform as she cut his grapefruit.

Mrs. Clinton supported the clemency decision when it was announced. But when it didn't play well in the press and the polls, she said she did not support it. Then, when Puerto Rican leaders in New York responded with outrage, she apologized for not consulting with them before denouncing the decision, and promised she would do so in the future.

None of this went over too well in New York, where it was almost universally assumed that Mrs. Clinton knew of her husband's decision in advance, may well have urged it, and was entirely prepared to announce that it was her idea if the decision had been better received.

There are other questions here, though, that have nothing to do with politics. Did it ever cross her mind, or his, to consider what the terrorists had done, and whom they had harmed, and how severely?

Did it ever occur to them to have compassion for, or protectiveness toward, the terrorists' victims?

These are important questions in part because the Clintons go to great pains to portray themselves as compassionate and concerned about all Americans. But in all the talk surrounding the clemency, the Clintons never spoke in any depth or with any feeling about the FALN's victims. There is absolutely no evidence that they felt any concern for them at all.

But those victims are worth considering here, for now they have been victimized again, this time in the name of political ambition.

Diana Berger was twenty-seven years old and six months pregnant with her first child when her husband, Alex, was killed in the FALN bombing of Manhattan's historic Fraunces Tavern on January 24, 1975. Alex was supposed to have lunch somewhere else that day, but the reservations got mixed up, and there he was just after noon, sitting with friends and colleagues at a table overlooking Broad Street when the bomb went off. Alex Berger was twenty-eight years old, a star graduate of the Wharton Business School. His only child, a son, was born three months after he died.

Diana Berger told the *New York Post*'s Andrea Peyser that she had spoken repeatedly to the White House after the clemency announcement, but all she received was arrogance. "I said, 'This organization has claimed responsibility for acts of violence.' He [a White House representative] replied, 'In this country we do not have guilt by association.' How insulting." She told Peyser, "How dare they say these terrorists have been punished enough? We've been punished each day, and will be forever."

Another man, Bill Newall, was with Alex Berger that day. Also at the table were Jim Gezork and Frank Connor. All but Newall were killed. "They died terrible deaths, barely recognizable to their families," Newall later told Congress. "It is impossible to adequately describe the effects of this savagery on the injured and the dead and their families." He said he didn't recall ever hearing, in the almost twenty-five years since that day, "any expression of remorse or concern or contrition by any members of the FALN for the pain and loss they caused unarmed and unsuspecting civilians."

Rocco Pascarella was injured in an FALN bombing on New Year's Eve 1982. A young officer with the NYPD, he was assigned to police headquarters downtown. At 9:30 P.M. they heard a tremendous explosion at headquarters. He thought it was fireworks. Soon they found out a bomb had exploded at 26 Federal Plaza, two blocks away. Pascarella was told to check out headquarters, walk around and see if he saw anything that might be an explosive. As he walked toward the rear entrance of the building he turned toward some debris. At that moment the bomb exploded. It blew off one of his legs below the knee, and badly injured the other. He lost half his hearing, some of his sight, required forty stitches in his face, and was in rehab for a year. After the president granted clemency, Pascarella testified on September 15, 1999, before the Senate Judiciary Committee. He said he couldn't understand why Clinton would grant any "clemency request with American blood on it."

Neither could Anthony Seft. He was a detective in the NYPD bomb squad that night, and was severely injured when one of the five bombs placed by the FALN exploded as he tried to disarm it. He told the House Committee on Government Reform that on that night he received "a lifelong sentence without the possibility of parole, with no time off for good behavior, and no chance for clemency. My sentence includes five reconstructive operations on my face, the loss of all my sight in one eye, 60 percent hearing loss in both ears. My only solace was that sixteen members of the FALN were serving prison sentences." He said the president "speaks out of both sides of his mouth," damning Timothy McVeigh's terrorism in Oklahoma but ignoring the FALN's terrorism now. He ended his testimony with the words "God bless America."

Would the Clintons really have found any legitimate reason to free the men who killed and injured these people? Of course not. Did they do it for mere political gain? Who doubts it? In America these days, at the end of the Clinton presidency, people no longer ask "What did the president say?" or "Did you hear Hillary's statement?" They say "What's their spin on that?" and "Did you hear how they're playing it?" Even their supporters talk like this. Their *friends* talk like this.

The Clintons always seem to believe they will get away with such scandals, and so far they have been proven right. The page turns, our attention shifts, new stories claim new headlines. And yet even among their supporters there is a looking away when these kinds of things are discussed, an uneasy shifting and a shrug. One senses with so many Democrats who support the Clintons an unspoken sentence: "They're all we have." They're the only big national winners the party has left. I always want to say to them: That's not true, you've got better than that, you've got good people, go look and you will find them. I think sometimes they're simply too tired.

The FALN story was big, and important. The Hillary-wants-a-house story isn't, but it is revealing.

Hillary wants a grand house in which to live, eventually, in New York. She finds one, with a big pool and a guest house, in high-bourgeois Chappaqua, in Westchester County, just north of Manhattan. Cost: $1.7 million. The Clintons can't afford it—Bill Clinton has just been ordered to pay a $90,000 fine, levied by a federal judge, for lying under oath in the Paula Jones case, on top of almost a million dollars he had to pay Jones herself, as well as the money owed, at least on paper, to all the lawyers who represented them in all the myriad scandals. But with a robust sense of what she deserves, Mrs. Clinton goes to a big Democratic fund-raiser, Terry McAuliffe, and asks him to personally guarantee a $1.3 million loan. McAuliffe, under investigation by the Labor Department, a man who dreamed up the idea of paying off party donors with a free night in the White House bedroom once occupied by Abraham Lincoln, says: You bet. Headline writers look askance: This, they say, seems . . . odd, and seamy.

Mrs. Clinton's aides announce she has checked the loan with the Government Ethics Office to see if it passes muster and has been told it does. This is soon revealed to be untrue: The ethics office had not vetted the loan's propriety; it had only ruled on what it called the "narrow" question of whether the loan must be reported as a gift.

A public interest group announces it will go to court to stop the McAuliffe loan. The Clintons soon announce that they just realized they don't need it. They'll get the house another way. And no, it has nothing to do with a public interest group going to court.

And so it goes.

V

Lying, of course, is not the Clintons' only distinguishing characteristic. They are marked, too, by an absence of grace, a lack of personal humility that is actually jarring, perhaps because it threatens to lower both standards and expectations for our leaders.

Think of the leaders of our times, and of times not too long past.

When World War II ended, Winston Churchill took to the balcony of Whitehall to speak to the huge crowd massed below. For the first evening since the war began the streets and buildings were bathed in light, and Churchill looked out at the roaring throng and said, "This is your victory." And almost as one the crowd answered, "No, it is yours."

When Lincoln, in April 1865, just weeks before the Civil War's end, journeyed to liberated Richmond, Virginia, the streets soon filled with newly freed slaves. An old black man who had never expected to know freedom took off his hat as Lincoln passed, and went to him and bent to kiss his feet. Old Lincoln

stopped him, brought him up full height, and told him no—
"Don't kneel to me . . . you must kneel to God only."

When Ronald Reagan was shot, as he recuperated in those
long weeks in the hospital, his aides one evening came upon him
on his hands and knees, cleaning the floor of his bathroom with
paper towels. He had gotten up to wash his face, he said, and
made a mess; water from the sink had splashed over the side,
and he didn't want the nurses to have to clean it up.

In the summer of 1999 I saw the pope, John Paul II, meet in
the Vatican with two dozen pilgrims from throughout the
world. As he entered—aged, bent, the Parkinsonian mask—a
delegation of Filipinos broke into cheers. He pointed at them
good-humoredly. "Philippines!" he said, as he raised his cane.
And then to some South Americans—"Brazil!" he called, and
they cheered because he was right. The pope walked on, and
then a young man in the robes of a priest or a seminarian, a
young Asian with thick tall black hair, put his hands in the position of prayer and bowed his head as the pope passed. The pope
turned toward him. "China!" he said. And the young man slid
to his knees and bent to kiss the pope's feet. But the pope put
out his arms, caught him, raised him as if to say, "No, I kiss
your feet," and embraced him as the small crowd broke into
applause.

These men had the humility of the great. It is a beautiful
thing to see or read of. It is a humility that says: "No, I am not
everything, I am nothing." This, in all of us, is the beginning of
wisdom.

But there is the humility of the non-great, too, the humility
of the uncelebrated suburban lady next door who seems to think

nothing of herself, who doesn't brag or bray and who keeps a family together and makes it all run. This is the humility of normal and yet elevated people who treat those around them with a consistent and sometimes wholly unmerited regard.

But the Clintons comport themselves as if respect is not what they owe you, but what you owe them. This is why they brag and praise themselves so much—to remind people of all they've been given by the Clintons. *"That's why for twenty-five years we've been working so hard in public service"* . . . *"Do you know how hard she works to make our country better?"* . . . *"Hillary is the most compassionate, the most hardworking, the most dedicated public servant I have ever seen in my lifetime"* . . . *"Bill has devoted his entire life to helping all of us"* . . . *"I have been concerned about and speaking about and leading in the area of children's issues for almost three decades . . . "*

Whatever the triumph, the people didn't produce it—they were given it by the Clintons. The White House press release never says, "The Dow hit 11,000 today and this is good news for America"; it says, "The Clinton Administration's economic policies have yielded yet another triumph." Any mention of the people who created the economic miracle is absent or pro forma, and in the last paragraph.

Throughout their administration they have betrayed this illness of ego. They have seemed not self-satisfied but self-saturated. And the point is not that they lack the humility of the great, but that they lack the humility of the normal, which is another matter.

* * *

VI

What is there to explain their behavior?

Over the past seven years countless observers have puzzled over the Clintons and tried to understand them, to find the piece of string that, once pulled, unravels to reveal their secret. We have watched them, talked to them, and pored over articles, books, and interviews looking for insights.

I offer here a small one, a suggestion of part of an answer.

I was reading a book when it all seemed to come together. It was *All Too Human*, George Stephanopoulos's memoir of his years as an aide to Bill Clinton, a book whose semi-candid tone and occasionally candid content nonetheless contains, as such books always do, revealing scenes. As I turned the pages the scenes seemed to come together and form a kind of mosaic. A mosaic is by its nature somewhat crude, and yet it does form a picture.

Here are the scenes:

Hillary Clinton, in a White House office, is disciplining her husband's staff when suddenly she collapses in a heap, sobbing, "I'm feeling very alone right now. Nobody is fighting for *me*." Stephanopoulos returns to his office and breaks down; "Fuck her," he rages. Soon he has audial hallucinations. Vice President Gore takes Stephanopoulos aside, and George understands the message he's being given: *"[I]f it comes down to you or me, I'll cut your nuts off."* Now Dick Morris is warning that "Ickes has something on [the president]." Now the president is on the phone with Senator Bob Kerrey, screaming, "Fuck you!" Soon a conservative wins an off-year election and Clinton is saying, "It's

Nazi time out there—we've got to hit back." Thinking about the man he helped put in the White House, Stephanopoulos judges him to be "mercurial," "weak," "prone to temper tantrums." Now the president is screaming—an "outburst," a "roar" marked by "purple rage" and "eruption of a resentment." Now, "worst of all," the president is doing "the silent scream, in which he silently rages and refuses to say why." A "pretty, busty, flirty" girl named Monica is roaming the halls asking if you'd like a double-tall latte . . .

Reading these scenes I suddenly thought: *These people are not quite stable. They're not completely mad, they don't wear tin foil hats and talk to chairs, but they are extreme in their actions and behaviors.*

And as I thought this, I remembered a pained face. It was the face of a tall, calm, and disciplined Secret Service man. I had first seen him in the Reagan era, and bumped into him once visiting the Bush White House. I didn't even know his name, but I knew his face, and knew how good he was at what he did. I hadn't seen him in years when I bumped into him at the 1996 Democratic National Convention, which I attended to write about the convention speeches for *Time* magazine. I saw him in a hotel lobby and said hello, and we chatted, and he seemed genuinely happy to see me, which surprised me because we hadn't been friends, only acquaintances. I asked him how things were going. And he stood there, and looked me in the eyes, and barely, just perceptibly, shook his head back and forth. As if he didn't have words; as if the words he had should not be spoken. We said nothing for three or five seconds. And then I said, "It's bad, isn't it."

"You have no idea," he said softly. "You wouldn't believe."

And then he said goodbye, and walked by himself through the lobby. And I wondered if seeing me hadn't simply reminded him of other, older White Houses, the ones he'd known before the current trauma, the ones that had given him his first and lasting sense of what a White House is, and how it operates.

We've never had a White House like the Clinton White House. I worked in Ronald Reagan's White House, had friends in and visited George Bush's, and visited a third, Jimmy Carter's, because a friend worked for him. Jimmy Carter's White House was staffed with normal, hardworking people. They had the usual assortment of human failings—pride, discord, resentment—but they were sane, well-meaning, and serious. Ronald Reagan's White House was high-functioning but riven by divisions, an intense place with the normal assortment of human flaws and foibles, plus a dash of hauteur. But it was run by stable and serious people. The wiggiest thing that happened in the Reagan White House occurred after the president was shot by John Hinckley, when Nancy Reagan began turning to astrologers to tell her what days were safe for the president to travel. George Bush's White House had the usual genteel jockeying and sycophancy, but again, they were serious, and nothing if not grounded.

But the Clinton White House, with its drama and volatility, with its seeming combination of ferocity and immaturity—there has never been one like it. The gargoyles, as someone once said in another context, have quite taken over the cathedral.

It is odd and unsettling to think that we, as a people, have grown used to all this. But to get a sense of how odd this White House is, take the Clintons out of the picture for a moment. Let's take one of their stories, the sex scandal of 1998, and stick to the known facts, but let's remove the Clintons from the story and make it the story of . . . George and Barbara Bush.

Imagine: It is 1995 and George Bush is in his second term as president. One day, we later find out in a government report, he is simultaneously being sexually serviced by a young intern in a hall off the Oval Office and talking on the phone about the war in Bosnia. The relationship between the president and the intern includes late night "telephone sex" on the president's private telephone line.

President Bush is informed that these calls—lengthy, graphic—are being listened to and presumably taped by foreign intelligence operatives. This opens the president, and the United States, to blackmail. A foreign government could now ruin the American president by making such tapes public—unless, of course, the president bows to the blackmailers and gives them what they want.

Does Mr. Bush stop having telephone sex with the intern? No. Why should he? He enjoys it. But he tells the intern that if they're ever questioned about the calls, they should say it was all just a joke—they knew they were being monitored, the phone sex was just a "put on." Then he continues calling her late at night.

At the same time, President Bush is being sued in court in a sexual harassment civil rights case. His relationship with the intern becomes known to the attorneys in the case. Sexual predators often repeat patterns; the attorneys want to question the intern. The story breaks first on the Internet, the newest and most unmediated of all media tom-toms, then on front pages across the country. Bush must choose: trust the country, or lie. He can turn to the country, speak honestly, and allow the facts of his imperfect life to be judged by a hundred million adults who are themselves imperfect, and most of them well aware of it. Or he can decide to "win this thing" by denying all. It might work, though it will likely drag the nation through a year of trauma.

He decides to lie. He sternly shakes his finger at the cameras and declares, indignantly, "I want you to listen to me. I did not have sexual relations with that woman, Ms. Smith."

President Bush's aides are instructed to go on TV and radio to bash his critics, to attack their credibility and their character.

Barbara Bush, pale with fury, goes on the *Today* show and accuses her husband's enemies of making it all up. She says this even though she had in the past hired private investigators to tell her what her husband does with women; she knows his sexual m.o. In spite of this, Mrs. Bush now looks at Matt Lauer and insists that her husband is the victim of "a vast left-wing conspiracy." Then she makes a threat: "Some folks are going to have a lot to answer for" when all this is over, she says. And with her cool and bitter visage you know that she is saying: *We will ruin them.*

The country does not know what to think—would the Bushes be so low? Would a president and his wife drag us through this if he weren't innocent? It must be a lie, a fiction created and spread by partisans.

But in time things happen, answers emerge. There is a dress with evidence. There is a bitter acknowledgment under oath. There is a bitter speech in which President Bush, after nine months of allowing and contributing to a great public trauma, admits that he did not tell the truth and blames the prosecutor who caught him.

Then he begins to make speeches in which he compares himself to heroes like Nelson Mandela and Mark McGuire. When this doesn't play well he calls in ministers and announces, head down, that he is a sinner. That evening he goes to an awards dinner at the White House where he laughs, hollers, pumps his fist, and says, "Barbara and I are just lapping this up!"

Mrs. Bush has her staff leak to the press that she's hurt and shocked by her husband. She makes no apologies for her part in the extended drama, no apologies for the threats and the accusations, nor for the smearing of the special prosecutor and the intern. No apologies for what the Bushes and their staff did, together. Why should she apologize? She's a victim.

Soon conflicting stories about who knew what when start coming from the White House staff. "The President's men," *Time* magazine reports, "were saying Mrs. Bush had known about the intern all along, the first lady's press secretary . . . was calling reporters and telling them no, she hadn't; he'd deceived her too." They even spin against each other!

All of this, including the *Time* magazine report, was, of course, what happened to and was done by the Clintons.

But what would we say if it had been the Bushes who had done all this? I think I know what we would say. We would say, "Get these bums out of here." We would say, "They are a disgrace, get them out of the White House." We would have impeached the president in the House and convicted him in the Senate—because his own party, disgusted by his behavior, would have turned against him.

And we would somehow all of us feel that a refusal to rise up and remove them would be on some level an indictment of us.

You say, "Well—the people didn't do that to the Clintons so I guess they don't feel as you do."

But I believe that many people in our country, since the shocks of 1998, carry within them a low-level, low-key, barely perceived sense of shame that we have let the Clintons continue to lead us. And I believe that this is one of the biggest things the Clintons, together, have done through this scandal, and others: They have left us with a barely recognized shame that has resulted in a barely noticed demoralization.

They lowered us when they asked us to rise to their support.

The Clintons held onto their high position but at a great price, and as usual with them someone else paid the bulk of it. For when we allowed them to continue as president and first lady, the world understood us to be saying, "We can accept these people as our leaders, because this is who we now are."

That was a bad thing to tell the world and a dangerous thing to tell America's enemies. Because it heartens them, it stiffens their resolve by telling them that we are as low as they want to think we are.

To those who wonder at the contempt the Clintons' critics feel for them—some of the answer is in this story.

CHAPTER THREE

WHEN PEOPLE ARGUE ABOUT HILLARY CLINTON, the anti-Hillary person usually says something like "What about all the scandals?" and the pro-Hillary person usually replies "That's all exaggerated by the press" or "That's him, not her."

But there have been many scandals large and small, real and not only alleged, involving Hillary Clinton. One is Filegate, in which it was discovered that early in the Clinton Administration White House staffers had illegally requisitioned, were illegally holding, were apparently illegally poring through and perhaps disseminating the personal information contained in roughly nine hundred raw FBI files on Republicans who had worked in the White House for Presidents Reagan and Bush. The files existed because the FBI as a matter of standard procedure had, over the years, done full field investigations on the Republicans to determine if they might pose a security risk in working for an American president. Such files contain everything from the worst thing your worst enemy might say about you to data and observations from your friends, testimony from your old landlord—*No, I don't think he drank much, although I saw him walk unsteadily a few times—he said he had an ear infection*—rumors, gossip, financial and court records, and much else.

The man who controlled the files during the period in question was the White House security chief, a former barroom bouncer and campaign dirty tricks operative named Craig Livingstone, whose employment, he had told co-workers, was arranged by Hillary Clinton. (If this is true, it was not her worst personnel decision; Gail Sheehy, in the biography *Hillary's Choice*, reports that Hillary, as a favor to heavyweight Democratic contributor Walter Kaye, recommended the hiring of an intern named Monica Lewinsky.)

Mrs. Clinton denied hiring Livingstone; she even said she didn't know who he was. This is surprising, because she apparently used him to track down leaks, and even to identify the body of Vince Foster the night he died. A White House intern later told congressional investigators that he overheard the first lady greeting Livingstone by name, and in a friendly manner, in the halls of the West Wing.

No one has so far succeeded in getting to the bottom of the FBI files case. The administration has portrayed it all as a bureaucratic snafu, a matter of miscommunication. Several lawsuits continue to make their way through the courts; news of depositions and affidavits continues to pop up in the papers, usually on page forty-three.

A White House computer specialist named Sheryl Hall said in court papers in December 1999 that Michelle Peterson of the

White House counsel's office told her that the administration's strategy was to "stall" the Filegate inquiry. Hall says Peterson told her that they "had just a couple of years to go." (The White House said Hall's accusation was "baseless.")

Hall said that while employed in the Clinton White House, she had been directed to take part in creating a database from personnel files. She said that she had refused to circumvent legal restrictions, and that a memo was then sent to Mrs. Clinton and presidential adviser Bruce Lindsey describing Hall as "disloyal." Soon her staff was cut, and she was relieved of any future responsibilities for the database.

Hall left the White House, and has now brought suit against Hillary Clinton and the Democratic National Committee. She says she was forced to quit her job because of "actions undertaken at the direction of Mrs. Clinton and in retaliation . . . for . . . challenging the unlawfulness" of the creation of a White House database from the files.

In January 1999 another former White House aide said in a sworn statement that she saw high-ranking Clinton Administration officials looking through the top-secret personnel files in 1993. Deborah Perroy, of the National Security Council support staff, said the documents included copies of FBI background reports on men and women who had worked for the White House during the Reagan and Bush administrations.

One day, Perroy says, she walked into an office and saw an NSC official and his assistant looking through the files, and keeping notes. They seemed surprised to see her, she said.

She resigned after the incident, ending more than five years of employment in the White House. She said in her 1999 affidavit that Clinton Administration officials then threatened "to come after me with false charges and allegations in order to smear my good name."

She has since learned that six months after she left the White House her own FBI file was requested by the White House Office of Personnel and Security—headed by Craig Livingstone—in spite of the fact that she no longer required access or a security clearance.

From Perroy's affidavit: "Based on my experience working for the Clinton administration, I believe my FBI file was obtained . . . in part because of fear I would divulge information about improper activities I witnessed at the Clinton White House, and so if I did go public . . . confidential information about me from my FBI file could be used against me." She told the White House she would sue if she were harassed.

Someday, probably after the Clinton Administration is over, all the facts of Filegate may emerge. Until then, it's another Unsolved Mystery of the Clinton era. Others involving Mrs. Clinton include the Rose Law Firm billing records, the facts surrounding the Whitewater land deal, and the still unresolved question of what was taken from Vince Foster's office the night he died.

But perhaps the most interesting is what happened in the White House travel office.

To understand what happened it's important first to understand Mrs. Clinton and the idea of "the story." When talking politics in private, Hillary Clinton often refers to what she calls the story—the narrative line the White House comes up with to create an impression in the public mind. Sometimes the story is used to illustrate a policy—the White House will use props and human interest stories to communicate why a bill should be opposed. Sometimes the story is a narrative conjured up to explain an occurrence that is otherwise inexplicable, or suspicious. After three years of claiming they could not be found, Mrs. Clinton's law firm billing records were suddenly discovered, she says, in a box on a table in the White House residence. The first lady said she has no idea where they'd been or how they got there, but suggested they might have been put there by an aide who had found them while cleaning.

Hillary Clinton likes the concept of the story because she learned it from political operatives whom she considers savvy. But she has also seen stories work. (Not with Hillary and her friends—they are too sophisticated to fall for the story—but with the trusting and distracted.)

Hillary's stories usually provide a motive, and the motive is always noble.

Stories are wonderful things, and the Clinton White House has told so many of them. But stories can be used to illustrate more than policy; stories can illustrate, even illumine, the character of the storyteller.

On May 19, 1993, four months into the first Clinton Administration, the seven men who had worked for decades in the White House travel office were suddenly and very publicly fired. These mid-level, nonpartisan government workers helped arrange and implement presidential trips, made sure journalists who traveled with the president had accommodations and could make their way through local customs. The travel office workers served at the pleasure of the president, but they had always kept their jobs through changing administrations because they were known to be good at what they do.

That is, until the Clintons came to town.

The man who ran the travel office was a veteran civil servant named Billy Dale who had gone to work in the White House in 1961, and who had run the travel office for more than twenty years. But when he and his staff were dismissed it was with what seemed a startling cruelty. The workers were charged with "gross mismanagement," told to clean out their desks quickly, and escorted from the White House, which announced they were the subject of a criminal investigation by the FBI.

It was a big story, front-page news, and the Clintons got more publicity from it than they wanted. And they wanted some, because Hillary was trying to tell a story.

The day of the firings there was a contentious White House news conference in which Clinton aide Dee Dee Myers took 105 questions about the decision. Veteran White House reporters who had worked with and relied on Dale and his staff for decades were skeptical.

Myers defended the firings by blasting the fired workers. "There have been, there's basically very shoddy accounting practices—mismanagement, a number of things—and in order to correct those, we thought it was advisable to take immediate action, and so the previous staff has been dismissed."

In the following days the White House was questioned in the press about the firings, and soon heavily criticized. So it changed its story. A week after the Myers news conference, President

Clinton went on CBS's *This Morning* to offer this explanation: "The bottom line is if we can run an office with three that they were taking seven to run, and we can save 25 percent off a trip because we have competitive bidding when they didn't have competitive bidding, the press saves money and the taxpayers save money. That was my only objection. If anything wrong was done, Mr. [Mack] McLarty [White House chief of staff] will correct it. This is a do-right deal, not a do-wrong deal. Let's not obscure what happened. We were trying to do the people's work with less money."

So it wasn't mismanagement that was in question; the firings had simply been a cost-saving measure.

Soon after, Clinton aide David Watkins offered a third explanation—the travel office shakeup was part of Vice President Gore's National Performance Review. But almost immediately the vice president's office said this was not true.

Soon there was a fourth explanation. Watkins now said the travel office firings were part of an attempt to meet a Clinton campaign pledge to cut the White House staff by 25 percent.

So it was just the president trying to keep his word, as always.

Then there was a fifth story—George Stephanopoulos said the travel office staffers hadn't been fired at all, they'd simply been placed on indefinite "administrative leave."

Why all the stories? Because the real one was not pretty.

The staff of the White House travel office was fired because Hillary Clinton wanted to give their jobs—their "slots," as she was quoted putting it in private conversation—to operatives loyal to the Clintons. She also apparently wanted to give the travel office's $12-million-a-year operation to a longtime friend and Clinton fund-raiser.

Harry Thomason, the Hollywood producer who, with his wife, the producer Linda Bloodworth-Thomason, has long been close friends with both Clintons (and had served as a major fund-raiser for the Clintons in 1992), had expressed interest in becoming involved in the White House travel operation before the first Clinton inauguration—and long before any "gross mismanagement" in that office could have been discovered.

Thomason had formed a charter travel company to provide travel services for the 1992 campaign; if the White House travel operation were privatized and put out to bid, someone would make a lot of money. As the man who ran the campaign travel operation, Thomason would be a leading contender.

According to evidence that later emerged, Harry Thomason and Hillary Clinton discussed a plan to remove the travel office workers during the transition period that followed the 1992 election. Thomason pressed his view that the travel office workers "should be replaced because they are disloyal." The quote comes from the notes of Hillary Clinton's chief of staff, Maggie Williams.

But Harry Thomason wasn't the only one pushing for a change in the travel office. David Watkins, an aide in the '92 campaign who joined the Clinton White House, was also keen to make changes. He brought into the White House as his aide twenty-five-year-old Catherine Cornelius, an attractive Arkansan variously described as a close friend and distant cousin of the president. Cornelius, described by reporter James B. Stewart in his book on the Clinton scandals, *Blood Sport*, as "dazzled by her proximity to power, full of a sense of her own importance," had worked in travel and scheduling in the '92 campaign. Now she came up with a plan to run the travel office herself.

In the first weeks of the Clinton presidency charges began to surface within the White House that the travel office was run by "a bunch of crooks" who had been "on the take for years." The quotes are from Cornelius's later testimony to investigators; she said she was quoting Harry Thomason.

Cornelius soon was assigned to the travel office, which many involved apparently thought was a good idea as there was some very Clintonian dissension between her and her co-workers. David Watkins's deputy later went on the record calling Cornelius "useless." Cornelius later went on the record saying that Watkins's office was "hostile to women" and that Watkins "touches people and pulls their hair and does stuff like that that was inappropriate." Watkins denied any misconduct. He was later fired by the White House for using a government helicopter to go to a golf course, a scandal that, as one observer said, may or may not have been a setup. (It perhaps should be noted here that every place where power is concentrated has the potential to become a shark tank, and the kind of free-floating power that exists in a White House, particularly a disordered and unstable one, does not tend to bring out the best in people. But the Clinton White House appears to be uniquely duplicitous and dangerous, like a shark tank in which the sharks are mad with hunger and high on amphetamines.)

When he assigned Cornelius to the travel office, Watkins told her to let him know if she observed anything untoward. Soon she was listening to private phone conversations, and secretly copying financial records and taking them home. She discovered that some staffers had "lavish lifestyles." One had a $6,000 boat, one had a modest vacation house, and one had recently been to Europe. She discovered, too, that the staffers used "racist" and "sexist" language.

Now Harry Thomason went to Hillary Clinton. In his sympathetic biography of the first lady, *The Seduction of Hillary Clinton*, David Brock writes, "Thomason must have known that Hillary, who had made it her business to oversee the operations side of the White House, was the person of action . . ." Hillary encouraged Thomason to move forward, and said that rooting out corruption in the travel office and implementing a new money-saving system would make a "good story."

But, as Brock noted, "Apparently, she did not stop to think that it would be a 'good story' only if the evidence [against the travel office] bore out the corruption charges . . . Hillary did not see that the far better press story would be how Clinton's friends, who stood to benefit, were behind the firings."

Thomason told Watkins that Hillary was ready to fire the travel office workers that day. Watkins went to White House counsel Vince Foster, the person closest to Hillary on the White House staff, and told him of Cornelius's suspicions. Soon the FBI was called in to investigate by White House counsel William Kennedy III, who, according to later testimony, told FBI agents that interest in an investigation of the travel office was being felt "at the highest levels." The FBI balked—there was insufficient evidence to launch an investigation. Two FBI agents, Tom Carl and Howard Apple, later said that Kennedy told them that if the FBI wouldn't investigate, he would call in the IRS. (Kennedy denies he said this.)

But Dennis Sculimbrene, the FBI liaison in the White House, backed up the two agents; he later said Kennedy had told him, "If the FBI doesn't do this, we'll have the IRS do it." Sculimbrene said that he was asked by senior Clinton staffers about the competence of the travel office staff, and whether he knew of any possible criminal activity. Sculimbrene said, "I told them at the time that they were very competent, that to my knowledge they were honest people. All seven had passed FBI security checks. And the press had never complained."

Vincent Foster had qualms. (One gathers that in his relationship with the Clintons, Foster was the Vice President in Charge of Qualms.) He suggested an audit of the travel office, yanking a group from the accounting firm KPMG Peat Marwick out of Vice President Gore's office and assigning them to the travel office over the weekend.

The surprise audit found some poorly documented records and $18,000 in unaccounted funds. The audit uncovered no evidence to support any other charges—no evidence that anyone there was "on the take," "disloyal," or living "lavish lifestyles." (Billy Dale later said that it was in the nature of his travel office duties that cash had to be kept on hand for unexpected expenditures, and that over the years it had become his habit to keep cash in various accounts to deal with various outlays.)

Chief of staff Mack McLarty brought the news of the audit results to the Clintons at a private dinner in the residence. McLarty's notes of the meeting include the words *HRC pressure.* The next day McLarty told Watkins that "immediate action must be taken." A day later Watkins sent McLarty a memo outlining a plan to fire the workers, with a copy to Hillary.

The firings were carried out two days later.

After Dee Dee Myers, in her White House briefing, announced the FBI investigation, the White House was accused in the press of using the FBI for political purposes. Now George Stephanopoulos convened a White House meeting in which he urged the FBI to issue a statement about the nature of its investigation. The FBI wanted only to say that the audit would be referred to the FBI for review. But Myers and Stephanopoulos had already told the press of a *criminal* investigation. The FBI agreed now to add a sentence saying that there was sufficient information for the FBI to determine that additional criminal investigation was warranted.

Before the firings, David Watkins spoke with the first lady. As Watkins's notes later showed, Hillary told him, "Harry says his people can run things better; save money, etc. And besides we need these people out—we need our people in. We need the slots."

According to Watkins, Hillary also told him she had been advised not to keep on holdovers from the previous administration. "She stated action needed to be taken immediately to be certain those not friendly to the administration were removed and replaced with trustworthy individuals."

It has been noted by others that in the weeks before the travel office firings, Mrs. Clinton had been angered by what she considered to be signs of disloyalty among the career White House staff. (Her suspicions were highly unusual; the Bushes had discovered no disloyalty, nor the Reagans, nor the Carters, nor the Fords.) One of the ushers, for instance, had talked to former first lady Barbara Bush on the telephone, and been fired by the Clinton White House shortly thereafter; one or more of the staff had gossiped about Hillary's arguments with the president, which resulted in press stories that she had thrown a lamp or an ashtray at him.

At any rate, Hillary early on evinced suspicions that people around her were leaking negative stories about the Clintons. She was also showing a tendency to divide those around her into two groups: My People and Other People.

Three days after the firings, the *Washington Post* headlined the fact that it had obtained copies of a memo from Thomason urging travel business be granted to his consulting firm. It also quoted from a Cornelius memo to Watkins urging that she be named to head the office and staff it with loyalists.

That's when the story became a scandal.

All along the way, as the travel office story built in the press and on the airwaves, Hillary Clinton and her staff insisted that the first lady had no knowledge of the firings, and then barely any knowledge of the firings, and then barely any involvement. In response to inquiries Hillary issued various denials, some under oath. She had "no role in the decision to terminate the [travel office] employees." She "did not know the origins of the decision." She did not "direct that any action be taken by anyone."

But years later the White House, under threat of a contempt of Congress order, finally released a long-sought-after block of 2,000 documents related to the scandal, and the personal, handwritten notes of her closest advisers and aides revealed that

Hillary Clinton had been deeply involved from the beginning.

Why? Again, consider the words of the otherwise sympathetic David Brock: "Hillary was caught in a web of cronyism."

Congressional investigators charged "political cronyism," and said "a pretext" for the firings was "created"; in the course, "the lives of seven innocent longtime employees were disrupted, and innocent citizens smeared." The *Washington Post* editorialized that the travel office workers "were dismissed for the shakiest of reasons . . . smeared . . . allegations concocted."

A tough but convincing judgment comes from former federal prosecutor Barbara Olson, in her book on the Hillary Clinton scandals, *Hell to Pay*. Noting that the president had the right to replace employees, she writes, "[The Clintons] did not have to smear them or lie about why they were being replaced. But it had become second nature to both. The abuse of power, the destruction of people who are in their way, the lying, even when the truth might be entirely acceptable—these are the Clintons' *modi operandi*."

It was later charged that White House lawyers altered documents that Congress had subpoenaed for its investigation. The *Washington Times* reported that the White House withheld from Congress chronologies and other documents concerning the first lady's role in the travel office scandal, and even altered listings of documents by deleting references to Mrs. Clinton in the documents' titles.

A few years later George Stephanopoulos, who was part of the group that pressed the FBI to say that it was investigating criminality—interestingly, in his 446-page memoir of his White House years he gives less than one page to the scandal, and the book excludes the key players he mentions from its index—was asked about the scandal by NBC's Tim Russert. Stephanopoulos said of Mrs. Clinton, "She's never denied knowing about it, but the question is did she order any action and she did not."

Russert asked, "Did she encourage it?"

Here Stephanopoulos spun the story in the classic Clinton mode: Throw up your hands in exasperation and implicitly accuse the questioner of *not understanding that this question has already been completely answered.*

"That's been gone over time and time again," he said. "There is nothing new there at all. This has been investigated by committees, been investigated by the White House and the role has been perfectly explained."

Actually it had never been explained by the White House truthfully, never mind perfectly.

A year and a half after he was fired, Billy Dale was indicted by a federal grand jury on charges of embezzlement.

On November 16, 1995, Dale was acquitted by a jury on all charges in less than two hours.

Bill Clinton was asked about it that afternoon. He said he was "very sorry about what Mr. Dale had to go through, and I wish him well. And I hope that now he'll be able to get on with his life and put this behind him."

Not quite "You better put some ice on that," but marvelous in its own way.

The president was at least more gracious than Hillary. After Dale's acquittal she was interviewed on *Today* by NBC's Maria

Shriver. "But did you want those people fired?" Shriver asked the first lady. "Did you think that was appropriate?"

Hillary Clinton answered, "Well you know, once the accounting firm found that there was financial mismanagement, the White House, I believe, acted the only way that it could have. Now there have been—"

Shriver interrupted: "By firing the people."

Mrs. Clinton responded, "Well, by—yes, by saying, you know, 'We have found evidence of this.' Now clearly we are the ones who said—namely the administration—that there were mistakes made in the way that the actual decision was implemented. Maybe it wasn't done quite as sensitively, and it turned out there was only really one person who was involved in this financial mismanagement. But the fact is, I don't know any American who cares about the integrity of the White House who would want anyone to turn a deaf ear to reports of financial mismanagement."

That's her story and she's sticking to it.

Hillary was playing hardball payback politics in the travel office scandal. She did in Arkansas, and she has in Washington. But because that facet of her nature is at odds with the caring and compassionate image she hopes to project, she tried to cover it up. Which is one of the things most disturbing about the affair: the disconnect between the reality of Hillary, the image of Hillary, and the ruthlessness with which she will go to any length to protect that image.

Did she care about the lives of those whose careers she was sacrificing to achieve her objectives? There is no sign of that in any of the notes or testimony of her aides.

Was she discomforted by the fact that she claims to stand for the little guy, and yet when some little guys got in her way she dispatched them with cruel efficiency? No sign of that in any of the notes or testimony of her aides.

Is there any evidence that she had qualms about smearing seven mid-level civil servants in the press, putting Billy Dale through an inquisition, and ending his honorable career in scandal? No. It was Vince Foster's job to feel qualms, and he soon killed himself. Which gave birth of course to other scandals, including the unsolved riddle of what papers were carried from his office the night he died. It is plausibly assumed by many that among them were notes and memos on Hillary and the travel office.

The FBI, too, was misused in the scandal, for it is obvious it was called in to provide cover and a rationale for the firings. This is chilling enough on its own, but even more so when one considers that years earlier, during the Nixon impeachment, a young lawyer working for Democratic investigators had helped draw up charges against Nixon and eloquently argued that he misused the FBI by ordering "investigations for purposes unrelated to national security, the enforcement of laws or any other lawful function of his office."

Hillary Rodham Clinton should have remembered those words, for she was the young lawyer who wrote them. And what she charged Nixon with is exactly what she and her staff did in what came to be called Travelgate.

In the end, the scandal was summed up aptly by Maureen Dowd in the *New York Times*. After David Watkins's notes became public, she wrote that Hillary's "maternal image" had been given "a Joan Crawford twist." "Mr. Watkins said he realized there was a more humane way to handle the situation than firing seven people, siccing the FBI on them, leaking it to the press, and pretty much ruining their lives." Dowd continued, "An associate of the first lady through all this confirms Mr. Watkins's portrait: 'She's a good screamer. She can cut someone to ribbons and make them feel like an idiot. It was a lot easier to do what she wanted.'"

But what is most disturbing of all is this:

Nothing that I have written here is new. All the facts are from articles, books, and news reports, many of which have been public for years now. And yet Mrs. Clinton somehow always manages to evade responsibility, and never to pay a serious price. Others do, but not she.

Her Teflon is tough; she gets tagged but it never takes. And a year or two after each scandal, when she is asked about it, she tells sympathetic interviewers that it's just so sad what has happened in politics, all the charges and the constant search for scandal. "It's a highly partisan atmosphere," she explains. And the stories go away.

But she doesn't do it alone. She does it with a corps of operatives and defenders, many of whose salaries are paid for by taxpayers, who make excuses, cover up, deflect blame, and who stigmatize and tarnish the reputations of those who do not excuse and cover up.

People say Bill Clinton has enablers, but so does Hillary. They are her staff, her friends; they work for her and have risen with her. In turn, she runs them ragged. She will call a staffer three times in a row in a restaurant as he tries to eat his dinner, even when he has told her beforehand that he is going out to dine with friends. But she speed-dials him anyway and he talks as he chews, and resents as he digests.

Hillary's enablers do what they do not because they love her but because they love their positions; because they hate her enemies more than they love justice; and because they are afraid of being targeted by Hillary's other operatives.

We'll end the story with Hillary Clinton's own words.

A few weeks before the travel office firings she gave a speech in which she declared, "We have to summon up what we believe is morally and ethically and spiritually correct, and do the best we can with God's guidance . . ." She encouraged all Americans to pay more attention to those who make society run but whose contributions are often overlooked. "You know, I'm going to start thanking the woman who cleans the rest room in the building that I work in," she declared. "I want to start seeing her as a human being."

A lot of people already see that person as a human being. A lot of people saw the travel office staff, those people who helped the White House run and whose contributions were often overlooked, as human beings. It's a pity the one woman in a position to so profoundly influence their lives saw them differently.

CHAPTER FOUR

E ARLY IN 2000 I SAT IN ON A MEETING WITH SOME politically active women who were interested in learning more about the political views of women who use the Internet. About two dozen women attended, almost all liberal activists of one kind or another.

Everyone was discussing what questions to include in an online poll, when an interesting thing happened. The talk turned to gun control, and soon it evolved into a discussion of the meaning of the Second Amendment to the U.S. Constitution. A conservative writer from Maryland said the amendment guarantees individuals the right to keep and bear arms; the question for us, she suggested, was to find out how women view various gun control initiatives. A Democratic activist cut her off: The Second Amendment allows not individuals but state government militias to be armed, that's what the Founders intended.

As they talked I had a thought. I had been wondering for

some time if the political left in America will eventually move to repeal the Second Amendment, perhaps in a new Constitutional Convention in the year 2010, or 2020. And so I suggested that we try to get a sense of support or opposition by including the entire amendment in a question, and asking women if they agree with it, would change it, repeal it, or leave it alone.

Silence for about a second. Then the Democratic activist furrowed her brow and said, No, no, bad idea. I asked why. "They won't understand," she said. "It'll be confusing, they won't get it." Everyone around her murmured in agreement: much too complex.

It seemed to me that they were alarmed by the idea of trying to find out what the people think—*because they might think the wrong thing.* And this in turn seemed suggestive to me of a certain paternalism in left-liberal precincts, as if they believe regular citizens can't figure things out on their own but must be led and guided in the right direction.

This is not an attitude of the old liberalism, but it is an attitude I see more and more, and I think of it in terms of a phrase that gained currency in the Clinton era—"command and control."

Command-and-control liberalism operates not in the open but behind the scenes, and not by airing issues but by handling them, or manipulating them.

It is command-and-control thinking that fatally shaped the creation of the Clinton health care plan, which was conceived and led, of course, by Hillary Clinton. It determined how that effort was carried out—in secret, and how it would end—in failure.

From beginning to end the health care plan was a perfect bureaucratic expression of Mrs. Clinton's leadership style, and New Yorkers might want to revisit the story before they choose her as a leader.

Throughout the 1992 presidential campaign, both Clintons promised that if they won the White House they would reform the American health care system. The promise was a standard part of Bill Clinton's stump speech, and was often mentioned by Mrs. Clinton in speeches and interviews. In an admiring profile in the *Washington Post* on October 30, 1992, reporter Donnie Radcliffe wrote, "Unlike traditional political wives of the past, [Hillary] has no misgivings about talking issues. She has never stopped being one of her husband's key advisors and when she wants to, she speaks with authority. She also intends to have her own issues." She quoted Mrs. Clinton as saying she was "particularly adamant" that health care be the highest priority. A week before, Bill Clinton had been on the stump in New Jersey, saying health care should be a "major, major, major factor" in the election. He promised a drastic overhaul of the system, including universal coverage for all and lower insurance rates.

A few days after his first inauguration he made good on his promise. President Clinton announced the creation of the President's Task Force on National Health Care Reform, and appointed Hillary to head it. "I am certain that in the coming months," he said, "the American people will learn, as the people of Arkansas did, just what a great First Lady they have."

Hillary's appointment aroused the predictable concerns: In

heading and directing an important government program she was taking on an extra-constitutional role, and there was no precedent. As a practical matter her leadership was problematic—if it didn't work she couldn't be fired, and in fact none of the usual penalties for failure could be applied. Her relationship with the president might cause complications in negotiations with Congress, and members of her own party might be inhibited if they wanted to support the president and his program but also wanted to complain that the Task Force chairwoman was making bad decisions.

Nevertheless, Mrs. Clinton proceeded as if she had been elected; and since the Clintons had run as "two for the price of one," they could claim, and did, that the American people implicitly supported her role.

When the president announced the Task Force, he said it would do its work in a way that was "as inclusive as possible." In retrospect it may be significant that his only specific suggestion on public involvement was that citizens write Mrs. Clinton at the White House and give her their ideas.

Certainly the Task Force had its work cut out for it. By 1993 health care spending made up close to 14 percent of the federal budget, and absorbed 14 percent of the gross national product—more than in any other country in the world. Within a few years it was projected to absorb an astounding 19 percent of the U.S. economy. And health care costs were continuing to rise, growing (as they had since the 1960s) faster than inflation.

But the American health care system isn't only expensive, it's huge, and complex. In 1994 it employed almost 11 million people, from nurses to doctors to dentists and druggists and technicians and support staff. There were 630,000 doctors alone,

5,600 acute care hospitals, and almost a million hospital beds. The system included every public, not-for-profit, and for-profit hospital in the country, and all the drug companies, insurance companies, makers of hospital equipment, and research and educational institutions.

And it all cost a trillion dollars a year. Government paid for 44 percent of it, insurance and managed care companies about 33 percent, and direct payment by the person treated accounted for 23 percent.

The system was complex, a real behemoth, and the first trick in "treating" it was to follow the wise old recommendation of Hippocrates, whose primary advice to new doctors was "First, do no harm."

Unfortunately, Hippocrates wasn't in on the working groups.

The first lady put presidential adviser and longtime Clinton friend Ira Magaziner, esteemed by the Clintons as a management whiz, in charge of supervising the Task Force. He created a brain trust of hundreds of experts and specialists; eventually there were six hundred people in thirty-eight subgroups and working groups, and soon chaos and confusion reigned.

So did secrecy. Task Force meetings were closed to the public and the media. This aroused anxiety among those who would be most affected by the Task Force's work, and the anxiety aroused antagonism. The Association of American Physicians and Surgeons went to court to open the meetings up. A federal judge ruled that Task Force working groups that were gathering data and formulating proposals had to follow legal requirements and stop meeting in private. In a key moment, the White House appealed. They *wanted* to meet in secret, away from the press and the people. This was a warning bell for potential critics. Soon Republicans in Congress found that they couldn't even get a list of the names and organizations of Task Force participants; the White House stonewalled each request for information, and Congress had to go through the General Accounting Office to get information.

Democratic congressional staffers had been asked to take part in the working groups, but Republican staffers were kept out, which was a matter of concern to them, and which aroused even more resentment. This is not at all inclusive, they said.

Soon the federal judge who had ruled on the working groups was threatening to hold the administration in contempt of court unless it started to produce documents, payroll records, meeting agendas and notes. He sharply accused the White House of giving only "dribbles and drabs of information at its convenience." (In the end, he ordered Hillary Clinton and others to pay the legal fees of those who had brought suit.)

Late in November 1993, the White House began to comply with requests for information.

Now, all of this happened within the first few months of the Task Force, and all of it was avoidable. But all of it was a reflection of Mrs. Clinton's leadership style—work in secret, fight attempts to learn what's happening, hold back information, and when people demand it, stonewall and delay.

None of it was necessary, and none of it helped the health care plan. The working groups and the whole Task Force not only could have worked in public, they would have gotten credit for having an open process on an issue so vital and personal to Americans. Moreover, the very complexity of what they were dealing with, when aired, would have helped them build a sympathetic understanding of the enormity of Mrs. Clinton's task, and the challenges she faced.

Instead, Hillary and Magaziner followed the command-and-control model: Don't tell people what's happening and don't let them in on deliberations. It's too complex, they won't understand, we know better.

Deadlines slipped, chaos reigned. Mrs. Clinton was out on the road, selling the public on a plan that did not yet exist, and back in Washington the brain trust was meeting, talking, trading data, and advancing options. Groups with a large and direct stake in reform were not represented but were sometimes briefed. Bureaucrats with agendas were running the show.

By the autumn of 1993, the main outlines of the plan were in place. Officials began showing pieces of it to Congress. Soon secret copies were making their way to the press.

Now the arguing changed from process to content.

The final plan was 1,342 pages long.

And what a plan it was.

Americans would have to choose among unfamiliar new "purchasing cooperatives" and "health alliances," and accept new limits on their freedom to choose doctors and hospitals; "gatekeepers" would determine who gets to see a specialist, and who the specialist might be. All employers—even small employers, local shop owners who employed a dozen people—would have to pay 80 percent of the cost of insurance premiums for workers and their families.

Questions and complaints were immediate. They were answered with details that heightened anxiety. When critics predicted long lines at HMOs, the rationing of health care, and second-rate service, the White House explained that that won't happen—we'll have doctors and HMOs concentrate on yearly performance reviews so we can monitor everything. People thought: Oh, give doctors and nurses more paperwork, that will make things better.

There was something for everyone to fear and resent.

It was soon clear that the plan was impossibly complex, hopelessly bureaucratic, and promised less accountability and higher costs down the road. It was marked from beginning to end by the presence of the heavy hand of government—and this in the 1990s, when a broad consensus had emerged that government is usually not capable of running programs fairly, economically, and with expertise.

The proposal managed to alienate almost every potential supporter. Big corporations that had started out as enthusiastic backers—their employee health care programs were extremely costly, and they would have been happy to see government take up the burden—were put off by new regulations. Insurance companies rebeled at budget caps. Small businesses said the requirement that they pay for health insurance would bankrupt them or force them to cut jobs. Doctors who had been for universal coverage were angered by their loss of autonomy in the plan. And deep in the proposal, hospitals found what appeared to be a plan to limit the number of medical specialists in order to limit the number of specialists' bills. That would be death for teaching hospitals. (New York, home of America's great teaching hospitals, take note.)

The plan was, finally, politically tin-eared from beginning to end. The big federal subsidies that would be required to create universal coverage and add new benefits would come, the administration explained, from big new taxes on alcohol and cigarettes, and big cuts in Medicare and Medicaid. Democrats and Republicans in Congress said: No way.

And it was completely *predictable* that they would say this. But then, Congress hadn't been consulted much in the creation of the plan. They might leak it. They might oppose it. They might apply old paradigm thinking. They might not understand it.

Until they saw the plan, the Clintons' allies in Congress had been able to maintain public support. But when it was submitted to Congress on October 27, 1993, support quickly fell apart. The administration failed to explain the bill or create coalitions and compromise.

Republicans comically diagrammed the proposal and produced a maze with arrows and boxes that made it look like the worst bureaucratic nightmare that had ever been created. Stalwart Democrats, like California's congressman Henry Waxman, said that the administration, in trying to sell the plan, seemed only to be relying on poll-tested jargon. "They introduced this very complex piece of legislation and then decided all they would say were phrases that had been market-tested, like 'health care security.'" The result, Waxman said, was that people began to think the administration was treating them "in a superficial way," as if they had "something to hide."

None of this was helped by word of some of Mrs. Clinton's reactions when pressed on the concerns of those whose businesses might be damaged by the plan. In her most famous comment, she said, "I can't worry about every undercapitalized third-party payer."

Democratic senator Pat Moynihan of New York, who had not, he felt, been treated with sufficient respect by the Clintons' staffs—in the early days of the Clinton White House an important administration official had told *Time* magazine that if Moynihan got in their way, "We'll roll right over him if we have to"—was now in charge of the Finance Committee, and in a position to roll right over the Clintons. He did. Instead of supporting the bill, he announced that there *was* no health care crisis. This suggested there was no need for a health care bill. Moynihan turned his attention to other issues; soon he called for an independent counsel in the growing Whitewater scandal.

There was one moment of promise. There were a number of competing health care proposals in both houses of Congress, but one, from Republican senator John Chafee, had the potential of wide bipartisan support. It was a scaled-back, more modest, less bureaucratic version of the Clinton bill. Chafee wanted to deal— but the administration refused. Hillary wanted sweeping change, nothing incremental; it was all or nothing.

She got nothing.

Different Senate committees fought over jurisdiction; the administration refused to choose among them; consideration was delayed; rival bills were created; the Democratic leadership and the administration refused to create a bipartisan bill. The administration was asked to compromise on various components; they refused. The Congress recessed, and returned knowing there was little support for the plan back home. What little momentum there was disappeared.

Hillarycare, as it came to be known, died.

The legal issues it had raised continued for a while in the courts. A judge asked the U.S. Attorney to consider prosecuting Ira

Magaziner for allegedly lying about the makeup of the Task Force; the U.S. Attorney said he did not believe he could make a perjury case, and blamed the White House counsel's office for using overly broad definitions of "members" and "participants" and "special government employees." The judge accepted the U.S. Attorney's findings, but said Magaziner and the government had made and maintained assertions that had gone beyond "strained interpretation" and were, simply, dishonest. Mrs. Clinton and others were fined $285,864. The White House said the taxpayers would pay it. A federal appeals court later knocked down the fines.

The political aftermath?

In November 1994, after the defeat of Hillary's health care plan, both houses of Congress were turned over to the Republicans for the first time in more than forty years.

Hillary Clinton lost her reputation as a savvy political player, and never headed a major governmental initiative again. On September 29, 1994, she told George Washington University medical students that she had not surrendered. "Health care reform," she said, "is a journey." Asked what she was doing now, she said she was reading Doris Kearns Goodwin's *No Ordinary Time: Franklin and Eleanor Roosevelt: The Home Front in World War II*.

Inside the White House they started forgetting to invite Ira Magaziner to important meetings; for a while he worked out of the White House on various projects. In 1998 he went back home to Providence, Rhode Island.

The GAO eventually toted up the price of the Task Force initiative—$13.4 million in preparing the health care plan,

$434,000 defending legal challenges to the way it was created. The White House—which had originally said the task force would cost less than $100,000—did not dispute the estimate.

And after all the Clintons' talking about and campaigning on health care, there were, in the year 2000, more Americans without health care insurance than there were the day the Clintons walked into the White House.

Mrs. Clinton, campaigning in New York, continues to insist that she will work better in Congress than the fiery Rudy Giuliani. But her leadership of the health care initiative is evidence that working well with others, and specifically Congress—working in a way that is truly collegial, and that shows respect for others—is not a strong suit of Mrs. Clinton's, but a weak one. It's not what she does, but what she doesn't do.

Again, New York, take note.

CHAPTER FIVE

I

SOME PEOPLE OPPOSE HILLARY CLINTON BECAUSE they think she has done so many things that aren't helpful, but I am often frustrated with her because she could do some real good, and at a crucial time, and doesn't.

She puts herself forward as a person who is concerned. It's a word she uses regularly in her weekly column: "I am concerned," "We should all be concerned." I just went to her website to see what she was concerned about today, and the very first thing I saw were these words: "Hillary Rodham Clinton: A champion for children and families. Hillary Rodham Clinton has been a leader in working to make life better for children and families, from her days in youth religious groups to her years as First Lady of the United States . . . "

But I cannot think of a time when she showed herself to be a champion by gambling her political capital; I cannot think of a single time in seven years that she jeopardized her position with her base to make progress for her country.

This is too bad, because as a political leader, as a candidate, and as first lady of the United States she has instant command of the nation's attention. She is respected by a number of people, including many of those who create our great newspapers and magazines, who run movie studios and produce television shows. Her followers are influential. They affect us all with their work every day. She could do a lot of good if she talked to them seriously about what they do.

I called this a crucial time. It is always a crucial time in America, of course. You can take a Nexus search back to any decade and see the quotes from columnists and politicians, "This is a crucial time . . ." Something big is always happening, and it always carries implications for the future.

But what is happening in America today is, I think, unparalleled in history. We are rising to the top and falling to the bottom; we are becoming wealthier as individuals and baser as a society; we are more powerful than ever and less mindful; we share a level of prosperity that is so high that it is a new thing in history, unprecedented in the story of man, and yet this wonderful thing we share has not made us closer as a people. And this has implications.

When I was in college in 1972 I spent a semester in England. There I read modern British literature, becoming particularly interested in the work of the playwright John Osborne, whose landmark *Look Back in Anger* famously changed the tone of British theater from one of decorous tea table tensions to charged working-class realities. At one point in the play Jimmy Porter, the central character, muses about the passing of all forms of English power and speaks of, "how sad to be anything else in the American age." It was a small aside in a long rant, and yet to me it seemed stunning—poignant, and full of implication.

He wrote that play in 1956. Now jump ahead four decades, to a Thursday night in May in 1999. I am attending a performance of David Hare's *Via Delarosa*, and listening, fascinated, as Hare, perhaps the most interesting English playwright since Osborne, muses on stage about his surprise at the flat banality of a Jewish settlement in Gaza. In his imagination it was a place of history, martial and dramatic. And yet, "It looked like the kind of suburban tract Steven Spielberg would have used to show America before the fall."

Again a stunning moment, and immediately I thought of Osborne—there's an arc for you, I thought, from wistful admiration to sad soft declaration. And one Hare knew the audience, composed of Americans from Manhattan and New Jersey and Long Island, would understand. (It did. There was a soft, soft sigh.) A few weeks before had been the events of Littleton, Colorado, and I think all of us in the audience were still unnerved by the fresh horror and the old sameness of it, for it had happened before. Columbine was just bigger, a dozen dead instead of three.

After the shootings I thought, as everyone did, about what was happening in our schools and our country. Everyone was asking who was responsible, and it seemed to me the obvious was true—the kids who did the shooting were responsible, but they came from a time and place and had been yielded forth by a culture.

What walked into Columbine High School in Colorado that day was the culture of death. It walked in wearing black raincoats; the time before, in the school in Paducah, it had been wearing children's hunting gear. Next time it will be some other costume, but it will still be the culture of death. That is the pope's phrase; it is how he describes the world we live in.

Think of it this way. Your ten-year-old child is an intelligent little fish. He swims in deep water. Waves of sound and sight, of thought and fact, come invisibly through that water, like radar; they go through him again and again, from this direction and that. The sound from the TV is a wave, and the sound from the radio; the headlines on the newsstand, the ad for the movie, all waves. The fish is bombarded. The waves contain words like this, which I'll limit to only one source, the news that you see on TV:

. . . *was found strangled and is believed to have been sexually molested . . . took the stand to say the killer was smiling the day the show aired . . . said the procedure is, in fact, legal infanticide . . . is thought to be connected to earlier sexual activity among teens . . . contains songs that call for dominating and imprisoning women . . . died of lethal injection . . . had threatened to kill her children . . . had asked Kevorkian for help in killing himself . . . protested the computer game, which they labeled sadistic . . .*

This is the ocean in which our children swim. This is the sound of our culture. It comes from all parts of our culture and reaches all parts of our culture, and all the people in it, which is everybody.

And we know this. It isn't news. It is part of the reason that Hollywood people, when discussing these matters, no longer say "If you don't like it, change the channel." They now realize something they didn't realize ten years ago: There is no channel to change to.

What "channel" tells our children that life is sacred—truly sacred, of inestimable value? Who tells them that they, and all people, are deserving of dignity and worthy of love?

Who counters the culture of death? The good parents and good families of our children, who are kind enough, sensitive enough, to give their children a sense that life has a purpose, a hidden coherence. Who are generous enough, and loyal enough to the future, to show through their actions that being kind is good, being responsible is good. "This is what we do," they do not say but show, "This is how we live."

But there aren't enough to go around! The children who wind up with guns in their hands often don't have anyone to counter the culture. Some parents feel bound down and defeated by it. And some simply forget to *see*.

We forget, those of us who are middle-aged, that we grew up in a time of saner images and sounds. For instance, the culture of crime only began to explode in the sixties. We have lived in it for thirty years, and most of us turned out okay. So we think our children will be all right, too. But they never had a normal culture against which to balance the newer, sicker one. They have no reference points to the old boring normality. We assume they know what we know: "This is not right." But why would they know that? The water in which they swim is the only water they've ever known.

This, I think so many people increasingly believe, is the great environmental issue of our time. We are one big continent with one big culture. It is one big ocean. The precious little fish that swim in it are endangered. And every thoughtful parent in the country knows it.

Soon after the shootings in Littleton, tornadoes tore through Oklahoma, causing vast and unusual damage. Whole blocks of houses, block after block, reduced to splinters, cars thrown on top of what once were roofs, middle-class and working-class parents and grandparents staring at the wreckage, concussed, and saying to the cameras, "I lost everything. I was just standing in the kitchen and the wind came and then—and now everything's gone."

There was an AP story a day or two later. I came across it surfing the web at night. It said that in the wake of the tornadoes, in the immediate aftermath of the high winds and turbulence, guns and bullets had dropped from the sky. People were starting to find them and turn them in to the police.

Let us suppose that God speaks in metaphor. What is God telling a country when he has bullets rain down on it from the sky?

I do not think I was the only person wondering this that night as I scrolled down.

We all know, we famously know, that things have gotten disturbing in America. And we all search for ways to make it better, and to make ourselves better.

II

It occurred to me after Columbine that there is someone who might help, someone who often speaks about children and who knows it takes a village to raise a child. Perhaps Hillary Clinton could come forward and speak about our culture in a serious and truthful way. It would take courage, because the people who produce our culture support her financially. But what better person to speak to them, friend to friend?

Soon she made a speech in which she decried not the culture of death but the culture of violence, as expressed in the growing number of violent video games and guns. We should think twice about these things, she said. Indeed we should. But that was all. In the next few days she was on to milk price supports, and how nice it is in upstate New York. The president called a meeting on violence in the media, and people came and said it wasn't good. Then the issue disappeared from the Clintons' attention.

I was disappointed. I thought Columbine might rattle Hillary into a new seriousness. I wanted her to take a chance and spend some of her capital, go out on a limb and say and do something of weight and import.

And then I witnessed something that you probably do not know about, something she did that astonished me because it really did take nerve.

A few months after Columbine Hillary asked to meet with all of the heads of the big TV and movie studios, all the media titans and moguls. She didn't tell them what it was about, and most of them thought she just wanted more campaign money and said they'd send a check, but her office called them back and said, "She wants to speak to you; she is coming to California; it is urgent." So they sighed and penciled it in. It's a myth that they love Hillary—they see through her and are often put off by what they see—but compared to the people who oppose her, those

Christian-right nuts, she looks good. And anyway, they see the world pretty much as she does.

The day came—it was the autumn of '99, after the new television season began. They met at the beautiful hillside home of Michael Eisner, the famous CEO of Disney. It was a mark of Hillary's enduring influence that he had, as she'd asked, invited his old foe Jeffrey Katzenberg of Dreamworks SKG. It was awkward at first, the cheerleader and the tip of his pom-pom, the head of Mouseshwitz and the little midget. But they acted as if nothing were amiss and chatted pleasantly, as enemies do. Hillary wanted Katzenberg there because she's stayed at his house and they're friends, and he's very generous.

David Geffen was there, and Rupert Murdoch, and Sumner Redstone, and Mel Karmazan came in from New York. Sherry Lansing of Paramount was there, too, beautiful and charming, and she hugged Amy Pascal of Columbia. Ron Meyer of Universal, Jeff Bewkes of HBO, Howard Stringer of Sony, friendly and witty. Edgar Bronfman who owns Universal came, as did Ted Turner and Gerry Levin of Time Warner.

Hillary arrived late, on Clinton time. She was very friendly but seemed breathless somehow; she kissed everyone and hugged Geffen and Katzenberg, and then Eisner to show she didn't take sides.

They all stood in the big sunken living room, with the sun dipping over the pool on the rolling lawn beyond. They could see it through the glass walls of the living room. It was beautiful, the Hollywood hills turning golden and deep.

And then, as if someone had passed the word, everyone sat on the couches and chairs, and Hillary stood before the main fireplace, and cleared her throat.

I should note here that I was there, too, miraculously enough with a tape recorder, so I can quote exactly. Why was I there? Because Michael Eisner's head housekeeper is one of my best friends—she is, to be exact, the first cousin of my best friend Lisa, and we all grew up together. I was visiting in the kitchen when everyone started to arrive, and because no one noticed me I just ambled toward the doorway and hit Record.

Hillary was speaking.

She said:

"I want to thank you all for coming here, and I want to thank you, too, for all the help you've given me and Bill. We owe you a lot. Without the financial contributions many of you make to us, reactionaries would move forward and take over this country. Together, so many of us have tried to stop them.

"But you've given me more than money. When Bill and I stay with Steven Spielberg and Kate in the Hamptons every year, we know we're getting more than a free house. We're getting a connection in the public mind with a man who is universally recognized as both good and gifted, the man who made *Schindler's List* and *Saving Private Ryan* and *E.T.*, some of the most appreciated entertainments of the entertainment age, and deservedly so. These films help us because they celebrate courage and love, and by celebrating, encourage them. Steven, when you have us as your guests you share your reputation for integrity with us. And we thank you for this, and we don't want to jeopardize that relationship, ever. We don't ever want to cause offense.

"But I want to be candid with you, for candor is a compliment, it speaks of trust and assumes good faith. And I'll never speak of this meeting and what is said, and I hope you never do, either. I want all of us to keep it private.

"We have a problem with our culture, as we all know—particularly those of us who are parents, and who have one night walked into the TV room and seen our kids glued to the tube watching something that is violent or darkly sexual or disturbed. Or if you've ever read the lyrics to the songs they're listening to, you know what's up. And you know we've all talked about this before, but I think it bears another look.

"And what I want to do tonight is have a conversation about where we are and who we are. I ask you to think along with me.

"Let me tell you my thoughts.

"First of all, I know what you know, and what any thoughtful person knows: Hollywood is always fingered as the culture's central culprit, but Hollywood is not The Culprit, it is A Culprit. If all of Hollywood started making shows with no violence and no obscene content and no bad messages tomorrow, the country would still be in a mess. There's plenty of blame to go around.

"But you know as I do that Hollywood is part of the problem. And a significant part. You make our movies and TV, and we are a nation that loves its movies and TV, that almost invents itself each day through what it sees and absorbs and internalizes from our media. You make the images that live in our minds, and prompt us . . .

"All of you went into show business to create wonderful shows and movies, to move people to laughter or tears or thought, and to do well in life through creating these things. I think one thing we all have in common here is that we were

ambitious when young. When you were twelve you were probably making Academy Award speeches in front of the mirror. I
was making inaugural addresses. Same thing. Dreams are good,
ambition is good; I honor these things, as do you.

"People who are cynical and uninformed say the only thing
that matters in Hollywood is money, but that's not true. I've
been in your houses, I've listened to you, and I know money is
not the biggest thing on your minds. What draws the greatest
part of your daily interest is the normal mix of human concerns
and anxieties—family and kids and workplace problems. And
after that what claims your interest is not money, but status.

"Status trumps money in Hollywood. Status—the position
of respect you hold in the community, your prestige. You want a
high place. This is understandable.

"Status trumps money in a lot of places, of course. It does in
politics. I've never been rich, but Bill and I have always been at
the top of wherever we lived, from Yale to Little Rock to
Washington. Everyone was proud to know us, and wanted to
stand next to us at the party, because at Yale everyone knew we
were going places and in Little Rock and Washington we had
arrived.

"If everything in Hollywood came down to money then TV
would be dominated by shows like *Seventh Heaven* and *Touched
by an Angel*, because these shows are very popular and make
huge profits. So if money were all, they'd dominate and set the
tone, which they don't.

"Where do you get status in Hollywood? There are a lot of
ingredients that go to the making of it—fame is part of it, and
money is part. But I think the primary ingredient is having a rep-

utation for fearlessness in creating your art—and getting critical acclaim for that fearlessness. You want to show in your work that you're daring and independent, that you bow to no norms, that you push the envelope and have no fear of those right-wing Christian fundamentalist know-nothings who condemn art.

"You gain status by producing *Sex and the City*, you lose it by producing or writing those nice little shows like *Promised Land*. They have no cachet, no one's going to introduce you at the party saying, 'Meet the fearless maverick who produces *Hallmark Hall of Fame*.' You want to be more experimental in your work, more realistic, more irreverent.

"What's another thing that would jeopardize your status? Well, taking the wrong side in the ongoing struggle over our culture. You would lose status if you were perceived by your peers to be going soft on your First Amendment rights to free speech. You'd lose status if you listened to those who support what you call censorship."

I have to stop quoting here to tell you something big happened at that moment. Harvey Weinstein lit a cigarette. Harvey doesn't smoke in the first lady's presence as a sign of respect—he knows she made the White House a smoke-free zone, and waves her hand as if you're polluting her airspace if you light up. So people don't. But Harvey, whose G-4 is nicknamed "The Flying Ashtray," all of a sudden lit a cigarette and started sucking on it like he was under water and it was his air tube. And everyone knew that this was not a mark of respect.

But it didn't faze her. She made believe she didn't see it.

*   *   *

She continued:

"And let's face it, we always call it censorship. But you know and I know that it is convenient for you to interpret it that way. If every attempt to critique is in fact an attempt to censor, if every attempt to persuade is an attempt to coerce, well, then, you always have complete license. You can do anything, produce the most vicious and sexualized and nihilistic material, even demonic material, and send it out into the country. Where, as we all know—let's just stipulate this—it does harm."

At this point there was another interruption, and it was a beaut. Ted Turner was over in the corner and he was starting to turn red and work his mouth in a kind of strange, jerky way. He was steamed. He stood up, and said, "Hillary!"

Everyone froze. They all call her Hillary in private, but in public or in front of people they always call her Mrs. Clinton.

"Hillary!"

She looked at him, and looked down, and looked at him again. She tried for a light tone. "Ted, I'm wondering if I could just finish my points here and then you—"

"Let me tell ya, Hill," he said, "I can see where this is going and I gotta agree. I've had a terrible goddamn problem this past year with obscene content on my news shows. I've had a terrible problem with the fact that to tell the news we've had to use phrases like 'oral sex' and 'telephone sex' and 'serviced in the doorway' and 'had relations as he took a call from a congressman.' With children watching! You say candor is a compliment, so you're about to get complimented: That was your husband's fault and your fault. He dirtied it up good, and you went along with it and dragged the country through it for a year."

Sherry Lansing looked down at her shoes, and Gerry Levin just put his hands on his knees and looked around as if to say, What is this? Eisner stepped forward and tried to intervene, but Hillary stopped him. Her face was red. She looked like she'd been smacked right in the face.

"Okay, Ted," she said.

And she continued:

"Duly noted. And I'll surprise you. Some of what you talk about was the fault of Ken Starr, who—"

"Who was asked by your husband's attorney general to investigate—"

"I know what Ken Starr did and who appointed him," she said.

She continued:

"I said I'd probably surprise you. I'm going to speak my heart—the real one, not the one I show on TV. I think Ken Starr was responsible for a lot of what went wrong, but let me tell you what I know now, and admit to myself. Bill did this. And I helped him. Because I thought it was right to defend him from people who are worse than he is. And I put on my game face and we fought, but we lowered everything around us when we did. Is that what you want to hear? Because it's what I think. And you know what? I'll go to my grave regretting what we did. I'll go to my grave regretting it. You know what else? It didn't strip me of my right to have a view and speak it. Because I'm an American, and no one can take that right away from you. Any more than they could take it from your wife, after she went to Vietnam and posed on the anti-aircraft gun and told them to shoot down more American planes. You know what happened that day? They tortured John McCain a little more that day because he wouldn't go out and meet with her.

"Michael Eisner, did you know that? You support McCain, you're his biggest supporter out here. Ask him what happened when Ted's wife Jane Fonda came to Hanoi.

"So I said there's enough blame to go around, and there's enough hypocrisy, too. And we all have to live with that. But we can't let it stop us from making progress."

It was a breathtaking moment. Ted was silent for a moment, and then he said, "Long as you compare yourself to Jane I know where you're coming from." No one seemed to know what that meant, including Ted. But it seemed to dissipate the tension, and people laughed, or at least made a breath-coming-out-of-the-mouth sound.

Hillary plowed on:

"Let's admit something else. It is interesting to me that you produce what you produce, and then take the greatest care to make sure that your children are protected from it. You buy your way out of the problem you help create. You send your children to the best, most traditional private schools, where they are insulated from the effects of your work. You have nannies who play with them on your rolling lawns, and make sure they're not watching too much TV, and if even with this they have problems down the road, you have access to the best thera-pists and advisers. So your kids are protected from what you do.

"And this is the great unspoken fact of America now, isn't it? The powerful—the politicians and producers and pundits, the people who write and edit *Time* and *Newsweek*, who produce the movies and talk about them on TV, the executive producers and writers and political operatives—they all support the free-dom to make any entertainment we like, they all support the cut-ting edge. But they make sure their own kids don't get cut and bleed from it.

"But what about the children of the powerless? What about the children of the kind of people we don't have to be with any-more, the unsuccessful? They aren't protected. Half the kids in our country are growing up in a sicker place, with parents too young or damaged or unsophisticated to protect them. They don't have the options of affluence. They're coming home from dead schools to dead neighborhoods, putting on the TV, eating junk food while they watch people get killed. And worse.

"Ted Turner, you and Gerry Levin own and run Time Warner, a great empire. You work hard, and you've brought up good kids. Gerry's son Jonathan was a hero, a young man who went into tough places to help needy kids, and he was killed by one of them. Gerry, you and Jonathan's mother must have done a lot of great things for him to help make him that kind of man.

"But I ask you to think with me for just a minute. Time Warner owns the cable system in New York, the state I hope to represent in the Senate. Channel 35 on your system is devoted, every night, to nothing but pornography that is utterly demean-ing of women, and to men. Think about this channel with me.

"All of our television channels are devoted to teaching our children what to want—a new car, or a job as a cop, or a house that smells fresher longer. Channel 35 teaches unparented and impressionable kids to want terrible things. It tells them that if you play your cards right you can use other people for your gratification, you can act out in a way that doesn't help anybody.

"And may I say that there are a lot of children in America who are deeply damaged, psychologically and emotionally. They have food to eat, they're not starving in the streets, but they are the new poor—they are the new face of poverty. They have not been given the emotional sustenance they need to reach a full and fruitful maturity; they haven't been given good discipline, and consistency; they haven't been guided by a strong father or a wise mother. Some of these children will remain immature; they will have poor impulse control, they will be incapable of patience, they will not achieve stability.

"There have always been such children, but now there are tens of millions of them, and we're making their lives worse by what we give them to live in. They are inspired by Channel 35, and by other things, Channel 35 is only one example among thousands.

"You know what's going on with music. It's not new; we've all been through the Marilyn Manson debate. But you know, it looks like those kids at Columbine were inspired by his lyrics that say 'go get a gun and kill.' And one of the most interesting things in that whole story is that Cassie Bernall, the girl they shot after they asked her if she believed in God and she said yes—we don't know if that story is true, but if it is—the fact is, that girl was into Marilyn Manson, too. But her parents intervened. They stopped her from going down the terrible path she was going down.

"Again—we don't have enough parents like that.

"So I'm asking you to step in. I'm asking you to think about these kids, the new poor, and to try and be sensitive about them and protective of them.

"Gerry, Ted, you could take Channel 35 off the air tomorrow. But there would be a big cost. Some of your colleagues, and people in the press, would say you'd bowed to censorship. 'Is Pat Robertson Programming Time Warner?' Your career would suffer; they'd call you a puritan, someone would look into your past and say, 'Who is this hypocrite to tell us what we can't see?' You'd lose status . . . and all for trying to help the kind of kids Jonathan Levin helped.

"But Gerry—I'm asking you to be as brave as your son was. David Geffen, you produce the music kids listen to. Howard Stringer, at Sony, you do, too. Be like Jonathan Levin, and try to help.

"Some of you are thinking, 'But I don't have a Channel 35, I don't put porn on the air.' But you there, Rupert Murdoch— you've actually admitted your situation comedies are probably harmful to our country. And this may be one of the few times you and I agree. I've watched *Just Shoot Me, Melrose, Dawson's,* all the sitcoms and dramas. I'm not kidding, I watch them. I can tell you the plot lines of *Felicity* and *Action,* and I've watched *Manchester Prep,* I think—the one they call 'a teen sex romp.' I know them all. A lot of these shows are bright and sophisticated. American TV is the brightest and most creative in the world, it's brilliant.

"But I also know what all of these shows teach our kids—that if you play the angles right, you, too, can grow up and have multiple premarital sex partners, glittering material goods, and a witty and cynical heart. A lot of these shows are real chilly, you know? They've got a chilly heart. They teach kids to be materialists, they teach them that the spiritual life is utterly absent from America and probably should be. Really, these shows are about sex and haircuts.

"They're not dramatically bad individually, but they have a cumulative effect—they're almost all that's on prime time. They're all our kids see. They won't end the culture, they just coarsen it every night. I won't even get into the fact that each season the writers and producers have to push the envelope a little more, get a little lower and coarser. Because you're all competing with each other. But where's it gonna end? How is *Sex in the City* going to be more demeaning—are they going to start dating German shepherds next year?

"And we have to start thinking about this, too: We're doing this in front of the world. This bears some thinking about. For more than a decade—since the mid-1980s really—American culture has been invading and claiming the world. Our movies and music, our TV shows, our *Eyewitness News* 'film at 11'—our ways, our attitudes, our perceptions—have swept the globe and entered the minds of children from Dublin to Tanganyika, from Capetown to Gdansk.

"The good news is that we have such sway—we own the world! We've successfully invaded every country! The bad news is that now they know who we are. They infer things from what they see. They used to see us—or, rather, those not influenced by the Marxist struggle used to see us—in those old movie terms: 'those good-hearted Yanks.' War in Europe? We'll send men. Famine in India? We'll send food.

"Do they see us as anything like that now? Or do they see us as the world's pornographers, as arrogant and mindless and insensitive people who endanger their children? And if that is how they are coming to see us, and I think it is, will they not someday take revenge? Or at the very least hope that something terrible happens to us?

"We have to think about this, because it's going to have implications down the road, for our children and their security.

"What does it mean when a great nation loses its reputation in the world?

"As you think about this, I am going to tell you what I want.

"I am asking you, as a supporter of your personal views and values, as a person who feels utterly at home with you and who has always defended your interests, I am asking you to change— to change utterly.

"I am asking you to stop making the entertainment that is hurting us, and start making challenging, tough-minded, truly fearless entertainment that doesn't rely on violence and sexual content to keep the attention of the audience, and that doesn't drum bad messages into the minds of children. I'm asking you to shun the producers and writers who make garbage. I'm asking you to do it privately, to privately come to agreement.

"You can do it, you have the power. People say you can't get the genie back in the bottle, but you know, you *are* the genie. And you own and run the bottle.

"And let's be frank: You're the only ones who can do it. We

talk about censorship, but nobody can censor you, because we have a First Amendment. If some idiot right-winger takes you to court you know the case will be thrown out, and you have the power to make sure he never wins another election and you just might use it. People can boycott you and write to sponsors, but boycotts don't work; they sputter out.

"Why? Because of another dirty little secret: Americans want what you do. They like to blame you for the mess we're in, but they encourage it every day. Give 'em more violence and sex, they'll love it. Because now they have a taste for it.

"No one can stop you but you.

"I'm asking for the greatest patriotic act of our time, but an act that is completely in line with your tradition. Because you don't come from pornographers, you come from the old men who used to run this town and who quietly came together one day and decided Nazism had to be stopped. Those old guys, Sam Goldwyn and Jack Warner—we make fun of them now, but they helped stop Hitler. Their war movies bucked up a nation, they stiffened the national resolve, they popularized sacrifice and cele-brated courage. And they didn't even know if there was box office in it. They just decided to do what's right.

"You can, too. Do it silently, so your enemies won't know and can't feel good about it. Do it now, so that a year from now when a mother in Knoxville and a papa in Brooklyn look up and say 'What's on?' and pick up the TV and movie listings, there'll be something for them to watch, something that for the first time in a long time will make them feel a little safe again . . .

"I truly believe that if you do this you will help save our

country. And you will become what they call your parents—you will become the greatest generation, the one that at some cost saved a world.

"And I'll do my part. I'll beat off the Christian right, I'll call them on their own hypocrisies, I'll go after your enemies, I'll humiliate Charlton Heston and the NRA, I'll stand by you. But you have to do this, in return, for your country.

"Let me end with two thoughts. I want to make it clear that I'm not saying we want a culture that is anodyne, flat, banal, and unrealistic. I'm not saying, 'Let's return to those good old days of *Donna Reed* and *My Three Sons*.' You know, those shows were so boring! Nothing ever happened in them, nothing interesting anyway. We make fun of them now, and we're right to.

"But this is worth thinking about. You don't have to be a genius to figure out that the guys who produced *Donna Reed* and *My Three Sons* didn't think they were depicting real life. They thought they were depicting an ideal—where everyone was kind and neat, where everyone ate at a table with napkins every night and talked and said, 'Meat loaf! Looks great, Mom!' How many families were like that? Mine wasn't, yours probably wasn't. But showing an ideal isn't bad. The kids who watched those shows got a kind of baseline for certain kinds of behavior: We were looking for clues on how to act, how to smile and give a compliment, how to be adorably smart-alecky, how to talk to Dad. We weren't just entertained, we learned.

"And children watching today learn things from TV, too. Don't imitate those old shows; you can't go back and we don't want to. Have more sophisticated content. Tackle big subjects, have fun and be witty and clever. But remember what Barbra

Streisand said in her song, 'Children will listen / children will learn . . . '

"And let me tell you something else. They say I'm a cynic, that it's all about me, that I don't believe in anything. But I actually do have some beliefs. And one of them is that America should continue. It should survive. And I don't believe it will unless we here in this room agree to change.

"And I've got to tell you, I don't mean to be rude and I am not making a threat, but if you don't change you have lost a friend. And I have been a good friend. But if nothing changes I'll try to make your lives difficult; I'll lead the charge; I'll lead a boycott, and this one you'll feel. Because if I turn on you, it's like Nixon to China: People will notice.

"I'd lose your support if I went down that route, and it might end my career. But some things are worth dying for, and I'm ready to die at the polls if that's what it takes to make progress.

"I do want you to know that I respect you, and I'm not saying that to be polite. You're brilliant, and you're leaders, and you're good people. And I want to wish you great good fortune as you move, I hope, toward greatness."

She waited for applause, as is her habit, but there was none. It was as if a fog of silence descended. And then they started to make noises and clear their throats and get up and move around. They thanked her for coming and for speaking to them, but they did it with a certain reserve. Katzenberg didn't hug her goodbye. Eisner got a laugh when he said, "I want to agree with the first lady on her eloquent stand on keeping this private."

And they all got in their cars and left, going to dinner at Eurochow, Morton's, and The Ivy. Hillary said goodbye, and got in her car, and she felt . . . different. She didn't know what she was feeling but it was a kind of combination of relief and pride, with a little anxiety around the edges.

She realized: This is what it feels like to tell the truth. This is what it feels like to take a chance to help our country. This is what it feels like to spend your capital.

She leaned back into the soft gray leather of the wide backseat and closed her eyes. "Let's go," she said to the driver. And she slept all the way to the airport, peaceful and spent.

Which is just when I awoke. Because it had all been a dream. It didn't happen. I had fallen asleep and my mind made it up— some antic little inner Edmund Morris took over my unconscious, and guided my hungry imaginings . . .

Oh, it was sad to wake up! I liked that dream. It was great to see Hillary being truly serious, truly earnest, and not just playing the part on TV.

It was so wonderful to see her focusing not on her career but her country. To see her jeopardize her position with her base and put her country's interests before her own, to see her act out of concern for Chelsea, whom she loves and whom she does not wish to see grow up in a base low place . . .

But of course it never happened, and really, it never will. Because Hillary Clinton doesn't feel she can put her country's needs before her need for campaign money; she doesn't feel she can put children before her career, for if she did she might lose her career. And then who would take care of the children, and the country?

And all of this represents another reason I think Mrs. Clinton should not be given any more power. Because somehow she never helps anybody with it but herself.

CHAPTER SIX

I

MRS. CLINTON HAS BEEN IN WHAT SHE ALWAYS calls public service and not politics for twenty-five years now. That is the record on which she bases the argument for her continued advancement: her long experience, her leadership, her proven ability to get things done.

But what has she done? In a quarter century of power both real and derived, what has she accomplished?

It is surprising to consider Hillary Clinton's record, because when you do, you realize with shock that she hasn't achieved much at all.

She has not created one program. She has not passed one bill, or lowered one tax. She has won not one legislative victory, not one electoral victory. For a very brief time she functioned on her own in the workplace, but became successful only when she joined the well-connected Rose Law Firm as the well-connected wife of the attorney general and then governor of Arkansas.

It can also be said (she has as much as said it herself) that she has not even lived a life of normal adult responsibility. She has for twenty of the past twenty-two years—that is, for almost all her adult life—lived in taxpayer-funded government housing. She has not had to make out a check for the mortgage or scramble to pay the rent; she has not had to pay the electric bill or shop for food or put gas in the car. The cars, most of her adult life, have been government sedans, driven by aides, state troopers, and later the Secret Service. (Bill Clinton would betray their detachment from the facts of everyday life when, in late 1994, he told his friend Sidney Blumenthal, who reported it in the *New Yorker*, that he had come to realize that crime was a big problem in America. This was 1994, more than a quarter century into the national crime explosion, the shards and shrapnel of which touched the life of every person in America—or rather every person who had not long been taken care of by a twelve-man security detail on round-the-clock call.)

Mrs. Clinton also had a profound advantage over the other working mothers with whom she sometimes competed in Arkansas's legal community. When she needed something at the last minute or needed something picked up, she had the state troopers do it. A state legislator had to threaten to launch an investigation before she stopped using the troopers to do her errands. And when Chelsea Clinton was born Hillary hired a nanny named Dessie Saunders, and put her on the state payroll as a security guard for $75 a week. The taxpayers of Arkansas paid for the governor and first lady of Arkansas's child care.

It is perplexing. She is so famous, so celebrated for her accomplishments; she so quickly and eagerly refers to them. And yet when you look at the record, her reputation for accomplishment seems to be just another fiction.

Hillary herself inadvertently confirmed some of this in her famous *Talk* magazine interview, in which she spoke of her Senate candidacy to the excited Lucinda Franks. "I want independence," she said. "I want to be judged on my own merits. Now for the first time I am making my own decisions."

I want independence? For the first time I am making my own decisions? How many people do you know who are still saying this of themselves at the age of fifty-two? How many women do you know who, at the age of fifty-two, have not already long lived lives of decision-making and responsibility? This is how nineteen-year-olds on the way to college talk, or twenty-three-year-olds entering the workforce. But most adults know well the land of adulthood and its demands, have grown used to bearing responsibility for the choices they've made; we're used to being judged on our merits. You don't really talk like this at the age of fifty-two unless you've been living a life in which other people do the heavy lifting.

Paul Greenberg, the columnist for the *Arkansas Democrat-Gazette* who has watched the Clintons up close for years, has described Mrs. Clinton's "royal" and "pampered" life. He has referred to "the uncanny resemblance" of Hillary Clinton to British royalty. "She derives her national prominence from her family position, not by election," Greenberg wrote. "Yet she manages to exude a proprietary air about her only-reflected glory, as if she were the one who had been elected. Her lifestyle is maintained at great public expense . . . She has run her home with the help of a fleet of maids, cooks, gardeners, chauffeurs, nannies and social secretaries, and yet presumes to advise the rest of us commoners on how to raise our children. When was the last time . . . that HRC had rushed the kids to daycare, worked an eight-hour shift, bought groceries on the return trip, folded laundry, cooked dinner, bathed the baby, and flopped into bed to worry about bills until falling asleep for five hours only to do it all over again the next day?" This life, Greenberg said, detaches the person who lives it from reality as it is experienced every day by regular rank-and-file citizens.

II

Her only great initiative in the White House was her attempt to reinvent the health care system in America. That was the only program she ever fully attempted, and she made both a scandal and a botch of it, creating an immense and impenetrable health care plan that even her most ardent defenders could not support, that even her party would not support.

Beyond the health care effort, Mrs. Clinton's career has essentially been a matter of talk and rise.

Talk by the Clintons about how much they care—talk that is Castro-esque in its length, that has quite filled the American air.

It is as if they confuse talking with doing, words with deeds, concrete accomplishment with chattering about aspirations. "Never confuse movement with action," Ernest Hemingway said. The Clintons are all movement, and the movement is all talk.

Part of this, of course, is generational. When the Clintons talk about history and how they experienced it, you realize that all their traumas were TV traumas. John Kennedy went to war; he fought for two years in the Pacific, came home, and, crippled with pain, went door-to-door in the triple-deckers of South Boston asking for votes as a veteran. Bill Clinton watched a war on TV and calls it his trauma. He watched the civil rights movement and calls it his proving ground. He—and she—were never immersed in history, in the midst of it; history was something that happened on a screen. And on the screen were images, and the images said things, and the things they said were enough. And so for her and him talk is action, words are action, and why not? You say them on TV and they're quoted, people repeat them, they have a reality in the American living room, so they must be reality, period.

When you're famous and you talk, the sound of your words goes on a spool of brown audiotape and is played in a microphone and is saved in the audio archives, and it lives forever. It is proof that you did something.

In the Clintons' case, the talk is self-referential—"Do you know how hard we've been working to help?"—and meant to show their hearts are in the right place. But one wonders if their hearts are even engaged.

And perhaps in the Clintons' case, their sense that talking is doing, that acting out compassion is compassion, that looking like a winner is winning, comes in part from something we all know, but that had a very specific impact on Bill and Hillary.

They grew up, these two exceptionally hungry people, surrounded by, inundated by, almost battered by images—images of the famous, images of the famous and loved, pictures in magazines and newspapers, in commemorative issues and movie magazines, on TV, on news shows, on interview shows, in the movie house, on the stand in the candy store.

The Clintons' generation was the first generation in history in which The Image was actually a huge and pervasive part of life. Pictures of others, of the famous, surrounded them like the air, and the pictures all had a narrative, the narrative of fame: *If you can get fame you will be happy and glamorous; everyone will love you, they'll want to be with you, children will reach for your hand. Adults will go weak-faced with awe in your presence.*

And that, for whatever reasons, seems to be what Bill Clinton and Hillary Rodham wanted. *They wanted to be in the picture.* They had hungry egos and they wanted to escape the boring, unimportant places in which they had been born and raised. They wanted to rise.

Everybody wants to rise, of course, but with the Clintons the need seems to have had a sharper, and ultimately destructive edge.

At the age of sixteen, as we all know, he decided he wanted to become president. The epiphany is said to have occurred on the summer afternoon when he met John F. Kennedy. We know about the moment because we've seen it, and we've seen it because young Bill Clinton stepped past the boys in front of him to shake Kennedy's hand . . . and get into the picture. (Look once again at that famous picture; Kennedy has a bemused look on his face, as if he'd seen this type before.) Which, Clinton could see, was being taken by the White House photographer over there on the right.

In Hillary's case, she was brought up by her mother to become, as she has told biographers and feature writers, the first woman Supreme Court justice. This is a great ambition, and ambition is good, and Hillary internalized it, and perhaps too well. She campaigned for awards that you were supposed to just quietly work for, a classmate who went to high school with Hillary told author Christopher Andersen. "It'll look good on my résumé," she said. The classmate was taken aback. "Nobody 'went after' it . . . you got the award because of your concern for your country, not so you could get the award.'" She won that award and others, and is seen smiling, holding them, in pictures in her yearbook. (We all become what we become, and few of us start out bad. But even young Hillary was consumed by ambition, and it was an ambition that changed with time, from "I will be helpful" to "I will be powerful," from "I will help put the spotlight on problems" to "I will put the spotlight on me.")

They were used to getting into the picture. The home Bill grew up in was remembered by his boyhood friends as a shrine to Bill; pictures of him and awards he'd won were all over the walls. If his mother was out having fun, she made it up to him by treating him, when she was there, like a young Viking god. Hillary's parents seem to have been made of sterner stuff—her father, from all that has been written, seems to have been gruff, and grumpy—but took a lively interest in the schooling and future of their elder child and only daughter.

Young Hillary and Bill, with growing hunger and perhaps some big emptiness inside them, latched on to the idea that if they got fame and power everything would be all right.

But they got fame and power, *and it wasn't all right.* In time they had simply traded one addiction for another, the need for fame and power became the need to continue having fame and power.

This is why they can't retire from "public service" when they leave the White House. This is why she is running for the Senate. This is why Bill Clinton can't stop campaigning—why his friends in Washington ask, when he's not in the room, why he can't stay in the Oval Office and govern. Why, they wonder, does he have to keep going out on the stump, going to every fund-raiser he can, even though he's not running for anything?

He needs to be the center of attention. Because when he is, he feels like he has a center. That's what the people on the rope line are there for: to fill him up.

Left to himself—left with himself—he is bored, anxious; he needs a hit of adulation or he will float away. (Clinton's aides dread drawing presidential vacation duty; Clinton can't stand being away from the limelight.)

She cannot live a life without power and admiration, or without the promise, the hope of power and admiration. (At first that promise resided in Washington, now it resides in New York.) If she did she would collapse and blow away, like an empty balloon.

He cannot live a genuinely private life. Neither can she. So they must be in ours.

I was thinking these things about the Clintons when I went to meet a woman who has known them for more than twenty years, a political professional who has been in the room and listened to Mrs. Clinton's conversation and watched her strategize. She had been inclined to like Hillary—they share the same political views and the same passion for politics—but as the years passed the woman found she couldn't. Still, like everyone who knows Hillary, she is fascinated by her, and thinks about her a lot, trying to understand her.

I told her I wanted to get her read on Mrs. Clinton, and added, in the hope that she'd tell me exactly what she thought; that she could speak for attribution or not. She chose not.

I asked her to bring her point of view, but she surprised me by coming to our meeting with a book, a classic of the field of psychology called *Borderline Conditions and Pathological Narcissism* by Otto Kernberg, M.D.

The key to understanding both Hillary and Bill Clinton, she began, is that they are narcissists. I said that this has been said before, and she answered that a big point has been missed: They are, together, a particular kind of narcissist, and they have done a particular kind of mind-meld.

She started to talk, and I listened. "One of the interesting things [Kernberg] says is that people with this kind of a disorder have a great need to be loved and admired by others, which is a curious contradiction between a very inflated concept of them-selves and an inordinate need for tribute from others."

She continued, "On the surface, they have an overinflated view of themselves, but yet they can't maintain that unless they get the adulation of an individual or, in her case and his case, a crowd—that is, the mirror that reflects back to them the grandiose view they have of themselves. So it's a circle that has to keep going."

She started to read from the book. "'Another characteristic of these personalities is that their emotional life is very shallow, they have little empathy for others, they have little enjoyment of life other than the tributes they receive from others or from their own grandiose fantasies, and they feel restless and bored when external glitter wears off and no new sources feed their self-regard . . . Their relationships with other people are clearly exploitative and sometimes parasitic. It is as if they feel they have the right to control and possess others, and to exploit them without guilt feelings, and behind a surface which very often is charming and engaging, one senses coldness and ruthlessness.'"

Is that them to you? I asked.

"Oh, yes," she said, "completely." She told me she'd gone to the book in her own attempt over the years to understand them, and what she'd read seemed almost a case study of the Clintons.

She continued reading.

"'Very often such patients are considered to be dependent because they need so much tribute and adoration from others, but on a deeper level they are completely unable really to depend on anybody because of their deep distrust and depreciation of others . . . This incapacity to depend on another person is a very crucial characteristic. These patients often admire some hero or outstanding individual and establish with such a person what on the surface looks like a dependent relationship. Yet they really experience themselves as part of that outstanding person; it regularly emerges in treatment that the admired individual is really an extension of themselves.'

"That is certainly true of both of them," she said.

I said, "You mean John Kennedy is his hero, and Eleanor is hers."

She shook her head. "It's with each other. Even with each

other, to the extent they admire each other, it's as an extension of themselves; that they see themselves as having the positive part of the other."

She said, "I think that . . . they are not capable of really one-to-one intimate relationships, whether it's with people they work with, or are related to, or people that they're close to all the time and . . . who do lots of things for them. Because that doesn't give them the feedback, that's not enough of a mirror. They need a big mirror. And if you think about some of the comments that were made about Hillary when there was first talk of her running in New York, someone who knew her said, 'Well, the question is does she want that small of a stage?' You know: 'She's global, she's not just the state of New York.'

"And I think that the bigger it gets, the more important the people are whom they surround themselves with, who will call them, invite them to something, let them use their house—if Steven Spielberg thinks they're exciting to be with, then they must be exciting to be with, because that reflection back from him confirms what they need. So if George Soros thinks she's brilliant, then she's brilliant. That's how they operate . . . They need that constant mirror image back."

It's not new to speak of the Clintons as narcissists—I suppose half the writers on politics in America have done that, including me—but this clinical description, coming from a thoughtful associate of many years, seemed both believable and somewhat chilling.

I think there's a broad sense, particularly in New York, where we all speak the language of therapy and have our Gail Sheehy moments, that the Clintons are psychologically complicated, to say the least. Half the people in New York think Hillary's candidacy is an acting-out of something.

But I'm not sure that it hurts her politically. For a lot of people it makes her seem more sympathetic. *The woman needs some help, don't be picking on her.*

Mrs. Clinton knows this, and plays to it; it's one of the things that makes her candidacy unique. Most candidates run for office saying "These are the issues, here's what I'll do to help you." Hillary Clinton is the first candidate to run saying "These are my issues—help me!"

Just before my conversation with the Clintons' friend I was reading Sally Bedell Smith's biography of Princess Diana, *Diana in Search of Herself*, and it seems to me that the Clintons are in that book, too. Diana was fifteen years younger than the Clintons, a full generation their junior, and yet the way Smith describes her made me think of them.

Smith writes, "Diana's preoccupation with celebrity meant she would never have a moment's peace. Had she decided to spend every day toiling in a shelter or hospice in East London . . . the photographers and reporters would have quickly disappeared. But Diana needed to alight, spread her magic, and move on. The magic might have withered if she had shed her glamorous mystique to pursue a life of quiet dedication. She also would have lost the ability to see her reflection each day in the press. The camera was kinder than the mirror."

Joe Klein made a point similar to Smith's, about Mrs. Clinton, in *Newsweek* a few years ago. He said that Mrs.

Clinton's claims of concern for the downtrodden would be more compelling if she had devoted some part of her life to really being with them, and helping them, with no fanfare.

Diana, according to many of the biographies devoted to her, followed the photos of and stories about her in the press with an absorption that seemed almost deranged. And when there were no pictures for a while, Diana went out and did something strange, or showed up somewhere looking beautiful, or gave a vengeful interview, to get back in.

Because only if she was in the papers was she real. The picture was as real as reality—realer than reality, for you can hold it in your hand.

And in thinking about all this I also remembered something I saw not long ago, an amazing scene I witnessed one night in the autumn of 1999 while watching CNN. The Clintons were in Kosovo, meeting with refugees from the Bosnian war. They were sitting and chatting with a refugee child. Television cameras were *burrrring* away, but then a print cameraman leaned down with his Nikon. Bill Clinton noticed him—I saw it. And at that point he moved his head close to the first lady's, so that their heads touched as they looked soulfully, together, at the poor children. It was— there is no other phrase for it—cringe-making.

There must be a strange paradox when you live a life so consumed by images. You want to be in the picture because it is real and will impress people, but the people you are impressing are by definition people who are impressed by a picture. They are not—how to put it?—insightful, bright, sophisticated. Which means you're spending all your time trying to impress people you think, in your heart, are dumb.

It must sour the soul.

Often when I watch the Clintons I think I perceive a profound joylessness, an almost glassy-eyed containment, or distance. It's as if they don't have a façade, they've become a façade. You sense a depression on his part and an anger on hers.

They seem to me trapped, lost in a maze of inauthenticity, looking for the admiration that they think will lead them home.

Two summers ago I bumped into an intimate, lifelong friend of the Clintons, and what he told me has remained with me as I watch Mrs. Clinton mount her campaign. The friend was on a plane with the Clintons in 1997, as they returned from a vacation. On the way back to Washington they had to stop for a ceremonial appearance. The plane landed, the ramp of Air Force One was lowered, and suddenly the Clintons, who moments before had been slumped distractedly in their seats, came alive. Now they were smiling, laughing, waving to the crowd in the lights. Suddenly the Clintons were happy.

"They're going to need a lot of therapy when this is over," he said. "They're gonna need help to get them off what they're hooked on."

What? I asked.

"The adulation," he said.

All of this would be touching if their hunger, their human if outsized hunger, could be met in some private way. But it has to be met, each day, and fed by our country. Which makes their private plight our public problem.

CHAPTER SEVEN

I

How exactly will Mrs. Clinton run in New York? What will she say, what general appeal will she make?

Let's think about the summer and fall of the year 2000, the campaign proper, if you will, before the voting in November.

She will be good on the stump, great on the rope line, and her most successful moments will be those choreographed in advance by her aides and strategists. There will, inevitably, be more tricky moments such as the FALN flap, but these will be limited if she resists (or is forced to resist) the temptation to micromanage the campaign.

She learns quickly and is disciplined, maintains focus, and appears to have high energy. She is quick on the uptake. Her staff will take care to make sure that she doesn't get too tired and the pressures don't become too great, for that is when phrases like "I could have stayed home and held teas and baked cookies," and "[I'm not] some little woman standing by my man like Tammy Wynette," and "vast right-wing conspiracy" pop from her mouth like tarts from the toaster.

When the stakes are high and the drama intense, her tensions rise and rub away the façade of equanimity, leaving her inner self exposed. That inner self is tough and aggressive. It was Hillary who whispered in the president's ear, giving him last-minute advice just moments before his furious "I want you to listen to me—I did not have sexual relations with that woman, Ms. Lewinsky" statement. On August 17, 1998, the day of the president's belated semi-confessional speech, it was Hillary who, worn by fury and eight months of scandal, ended internal White House debate by snapping to the president, "It's your speech, say what you want." He did, and was roundly slammed for a speech so graceless and lacking in compassion for his primary victim, the country, that he had to spend the next four weeks insisting on his remorse.

That kind of drama, the Clintonian drama of inflammatory language and reflexive aggression, may scar Mrs. Clinton's candidacy—but I don't think so. She knows, and has no doubt been informed by her aides, who have pored over focus group verbatims and poll results, that she cannot afford such behavior because it will underscore existing negative public perceptions of her.

With a consistency that would be called heroic if it were applied to a high purpose, she will keep her smile on, hug children, speak in airy phrases that no one could disagree with

("Our children are our future"), stick to the script, and try very hard not to answer questions. If cornered she will respond with pleasantly robotic answers of such banality and length that TV reporters in the edit room will choose not to use them because they will only slow the piece and confuse the audience. (When she wants to be quoted, on the other hand, her answers will be short and sharp.)

I suspect we will be seeing more of a small and subtle habit Mrs. Clinton has developed in the past year or so. When she is asked a question that touches on what she considers a danger area, and to which she wants to give a "no" response—"No, Dan, I don't think the American people are persuaded by all these highly partisan charges"—she often nods her head up and down, as someone would when they're saying yes. When she is asked a danger-area question to which she wants to respond with a "yes"—"Yes, Dan, I do think the people are responding to my message"—she often shakes her head side to side, as people do when they mean no. It is possible that this is an unconscious quirk, but I suspect it is deliberate, a way of taking the clarity out of a moment and leaching it of impact.

The conventional wisdom a year ago was that the famously ferocious New York press corps would be a major challenge and even a stumbling block for Mrs. Clinton. This may not happen. While not in the tank for Hillary, the press corps will likely engage in subtle and unspoken self-censorship. The general feeling of producers and editors about the first lady's race was captured in the summer of 1999 by the playwright Wendy Wasserstein, who wrote in a *Washington Post* op-ed piece of a recent dinner party of high-ranking members of the New York media. The guests spoke admiringly of Mrs. Clinton and supportively of her candidacy, but Wasserstein, a liberal Democrat, had reservations about what she saw as a Democratic Party "coronation" and demurred; in the silence she felt she had broken a taboo. After she wrote critically of Mrs. Clinton in the *New York Times*, she was upbraided at a script meeting by the powerful Harvey Weinstein of Miramax, a major contributor to Mrs. Clinton, who shook his head at Wasserstein and said, "I thought you were a nice girl."

The day Mrs. Clinton announced the formation of her "exploratory committee," in May 1999, at Senator Moynihan's upstate farm in Pindars Corners, New York, Mrs. Clinton's remarks were broadcast live on three news networks, and the press showed great enthusiasm. CNN's Bruce Morton said of the day, and of Mrs. Clinton's announcement of her upcoming "listening tour," "It's really the voters' show . . . the point is to hear what they have to say." But that was not the point, as surely he knew. MSNBC's Chris Jansing could barely contain herself. She made a sad face as reporters waited for Mrs. Clinton's appearance, and said of her candidacy, "It all began in the depths of her depression . . ." She asked reporter Chip Reid how Hillary was feeling in the moments before her remarks—excited, nervous? Reid, reporting from the White House, said he thought she was excited. "She's running on Clinton time," a now-happy Jansing said as Mrs. Clinton arrived.

In a piece the next day in the *Wall Street Journal* I called the appearance at Senator Moynihan's farm "a triumph of spin." Two days later a top aide to Mrs. Clinton took one of her friends aside. "I heard it was wonderful," she said. By wonderful, it was clear, she meant supportive of Mrs. Clinton. The friend said no, the piece had called the day a triumph of spin; it wasn't a compliment. But Hillary's aide still thought it was.

Why wouldn't she? Hillary Clinton's staff, like her husband's, think spinning is a high calling. They have been celebrated for it in the media. They have forgotten that when someone calls you a liar they are not praising you.

Few in the elite media, the networks and big stations and national magazines and big newspapers, will press Mrs. Clinton on the allegations of scandal that have marked her time in the White House. They will congratulate themselves on avoiding "sleaze" and "innuendo." In return, they will get exclusive interviews, but not many; Mrs. Clinton will avoid interviews as much as she can because she doesn't like uncontrolled environments. Some intrepid reporters—the *New York Post*, the *New York Press*, maybe the *Daily News*, Jimmy Breslin and Jack Newfield—will attempt to break away from the pack and past the Secret Service cordon to ask questions, but they will be reduced to shouting seemingly rude challenges as she hurries by. Reporters who press too hard will be called hecklers and harassers, and will be informed by colleagues over drinks that they are hurting their careers with their "antagonism" and "lack of objectivity." NBC, CBS, and ABC will nonetheless do pieces with names like "Hillary Faces the Gauntlet," taking a wry look at the famously irreverent New York press. Cut to the guy yelling, "How did you make that hundred grand?" and then a close-up of Hillary being interviewed by a sympathetic anchor. She'll say she understands how passionate people get about politics, but that the process has gotten uglier, which is sad. She will seem both patient and pained.

In any case, questions on scandals will get a mantra-like response—"I'm here to meet with all New Yorkers, and I look forward to working with them on the challenges that face us now and in the future." The implicit rebuke—*How can you be down in the gutter when we're trying to figure out health care?*—and its constant repetition will discourage persistence.

Until the campaign proper begins, late in the summer of 2000, Mrs. Clinton's strategy will be to establish her presence and nail down her base. Her tactic will be, in the words of a reporter who follows her on the stump, "to bore us into submission" with endless repetition and photo ops. She wants to be an almost daily presence in New York, with a minimum of news flaps.

From the first days of her campaign, the tactic of repetition, presence, repetition was clear.

First the slowly accelerating number of trips that she took from the White House to New York, each visit an event, with pictures—Hillary meeting with local mayors, Hillary at Jones Beach. Early in a campaign everything is a blur on a TV screen; she knew the blur-impression she was creating was that *Mrs. Clinton is here a lot.* In another three months it would be *Hillary kind of lives here . . . she must have an apartment here . . . she sure cares about us.* By the spring of 2000, some potential New York voters will think she's lived here for years and is running for reelection.

In considering the rest of Mrs. Clinton's campaign, it is useful to remember the observation ascribed variously to Nora Ephron and Lily Tomlin: "No matter how cynical I get I just can't keep up."

Her aides will tell her to act out humility. "Be modest on the stump—admit you don't know everything. That's how to put it, 'everything.'"

Her vinegary and experienced aides will fight like tigers. Harold Ickes, abused in the past by the Clintons, has what appears to be battered-wife syndrome—"I can't leave, he needs me!"—but is a serious ideologue who makes serious use of derived power. Mr. Carville of Louisiana will be on the scene in part to guarantee the goodwill of the newly elected prime minister of Israel, Mr. Barak, whose victorious campaign Carville was involved in last year.

Watch for Mrs. Clinton to come out in support of presidential clemency for Israeli spy Jonathan Pollard, serving a life sentence in prison in the United States for spying on the U.S. There is strong support in Israel for clemency, and among New York's Jewish community, which is an important part of Mrs. Clinton's Democratic base. Freeing Pollard would cause a national uproar but would likely play well in New York City, where Mrs. Clinton has to win 65 percent of the vote to counter expected strong Republican numbers upstate and in the suburbs.

A good bet: Mrs. Clinton will issue a strong call for clemency, and the president, after due consideration, will refuse. That would be what Mr. Clinton calls a twofer: She'll get local credit for trying; he gets national credit for resisting. Actually, it will be a threefer: It will "prove" they don't coordinate their actions. (Clinton in August 2000, to the press corps: "I told you last year during the FALN dispute, I have to protect the national interest, and it wouldn't be right to check this or vet this with the first lady.")

Because Mrs. Clinton is loyal not to place or people but only to her position she will easily make the leap from patronizing Hot Springs to patronizing Staten Island. She will hide her condescension well, and speak to the locals with warmth and humor. The old good-natured crowing about how Arkansas has the biggest watermelons in the country will become new good-natured crowing about the best cannoli in the world.

Republicans could not, in 1999, blast her positions in advance because her pollsters had not yet told her what they were. By the end of 1999 they had made it clear that Mrs. Clinton would soon begin to campaign as New York's Great Protector against the Republican Congress and its continuing radical impulses. The strategy has a famous name: triangulation. It is not clear if Mrs. Clinton will join her husband and slam Republicans who disagree with her as "racists." Her decision will be determined by the polls, and by how well the tactic plays in other races.

To nail down her base Mrs. Clinton will support liberal nostrums for social problems and align herself with all unions. She will not, however, refer to or support some of the more liberal positions she has supported in the past. As Michael Barone has noted in *U.S. News & World Report*, "When Hillary Clinton was chairman of the Legal Services Corp., its affiliates brought lawsuits to force New York's Transit Authority to hire heroin users, and to require racial quotas in school suspensions in Newburgh, N.Y." By the summer of 2000, having energized her natural supporters, she will become more pragmatic and campaign as a moderate.

She will stand with police union members on the steps of

medium-sized city halls and deliver speeches in which she supports the death penalty, community policing, and a ban on cop-killer bullets. She will accuse Republicans of arming criminals by refusing to back her husband's gun control proposals.

She will stand with nurses' union members under banners reading "Better Health Care for All" and announce, "We have to fight so that the hardworking people of New York aren't bankrupted by hospitals and ignored by HMOs." There will be few specifics.

She will spend a lot of the campaign in grade schools, reading to children, and finally speaking with indignation about how "the forces of cynicism and shortsightedness" want to "take money from our public schools to finance their experiments"— vouchers, charter schools that are truly independent—"and leave our new citizens, our immigrants, and those who have been left behind in a brutally broken system."

In 1999, Mrs. Clinton was frequently criticized by newspaper columnists in New York, and never responded. She seemed not to take it seriously; it did not deter her. It didn't deter her not only because it was inevitable but because it was at that point welcome.

What she was doing was absorbing attacks. She was playing rope-a-dope, exhausting her foes by taking every blow they could throw and allowing their charges to enter the air and turn into clichés—the kind of clichés that people eventually stop hearing.

By the fall of 2000, when it matters more what is said of her, her critics will look obsessed and belligerent. She will look long-suffering and patient, like a victim.

II

There is the matter of her speeches. She has big ones coming up, after her announcement. There will be big rallies in the fall, and they will be heavily publicized and covered. The more successful of them may be cut up and used in radio and TV commercials.

What is she thinking about her acceptance speech, the one she will give after she wins her party's nomination?

What is she thinking as she enters 2000?

Let us speculate.

She is thinking first of all that the people who say she's been daring in gambling her entire political career on a race she may lose just don't get it. Sure it takes guts, but the odds are with her for a win, if not a literal one then a symbolic one.

She is thinking that she has a 66 2/3 percent chance of coming out of this a winner. That is, the odds are two out of three that she'll benefit from the race. Those are the kinds of odds any gambler would take.

First, she may win. If she does she's a winner, the new U.S. senator from New York. But she's even more than that: She's Hillary Clinton, Feminist Icon, in the U.S. Senate. Every move she makes, every step she takes they'll be lauding her. From day one, there will be speculation: When will Hillary announce for the presidency? There will be background stories, with quotes from her Senate staffers: "She'll watch how President W does; if he's looking good in 2002 she'll hold back, but if he's looking vulnerable she'll start visiting New Hampshire and then start an exploratory committee early in '03."

There will be press stories about Hillary's long climb, what she's learned in the Senate, how she's mellowed. There will be comments from her fellow senators: "I thought she'd be a monster to work with but she's great, very collegial, and a sense of humor—she cracks us up in committee meetings." (That will be Senator Orrin Hatch.)

Little by little the scandals, the questions, the financial dealings will recede into the past.

So: If she wins she wins, and is the future Democratic nominee for president. She believes there's no one else in the party. Gray Davis is too gray; no one in the Senate has any national power or strength. She believes Gore will likely have had his chance and blown it, and by 2004 Bradley will be so—so 2000.

What if she doesn't win? She thinks she can still win.

If she runs a credible, respectable, intelligent campaign—if she makes some mistakes but is a good candidate, *and comes close*—well, then she's also a winner, and still has a political future. She will give a graceful concession speech. In time she will come forward and grant an exclusive interview to Tom Brokaw or Peter Jennings and say something like: "You know, I was born and bred in Illinois, and when you come from there there's only one man in politics you really learn from and get to know inside out. He had some early success in politics, and then he ran a valiant campaign for the Senate, and lost. And he lost to a lesser person. But only two years after that loss he ran for the presidency—and won. So I think Abe Lincoln is an inspiration to us, and he's been an inspiration to me all my life, and just as he had a future I hope I have one, too."

That will mean either she's running for Republican Peter Fitzgerald's Senate seat in Illinois in 2004, or she will go on to work for children's groups, making speeches, going to fund-raisers, appearing on television, keeping her contacts and support up within the party, and biding her time for another run—maybe for the presidency.

And then there's the third chance: the chance that she will lose in New York, and badly. She won't come close, she'll run a bad campaign and belly flop big time, showing she doesn't have it in her to be a candidate. But that, she is convinced, is not going to happen.

So while some think she's gutsy, or unrealistic, she thinks in her heart and head that in fact she can't lose.

III

What else is she thinking?

Let's imagine what she's thinking as she sits back, eyes closed, in her seat on Air Force One as she flies into LaGuardia for a campaign day . . .

Let's say it's April 2000 . . .

Let's say she's talking to herself, free associating about where she is and what's going on around her . . .

Maybe she's thinking about that offer from WABC Radio to go on the air once a week for an hour, like her expected opponent, Rudy Giuliani . . .

She's thinking:

I'll string out the ABC offer long as I can, good publicity, give the impression I'm pondering it. As if. It's a call-in show, journalism without the gatekeepers—the network news may kiss my hand, but Al from Bayside will kick my ass. No way. Mandy said she'd screen the calls, but screeners can get fooled. And anyway, when she screens out all the wingnuts there'll be no one to talk to except some "I'm Howard Stern's penis" person. Or even worse, "I'm Howard Stern." We'll string it out long as we can, then say I'm too busy meeting with people throughout the state to commit . . .

Got to think about the acceptance speech when I'm nominated at the state convention. Maybe Bob Shrum will help me, or Begala. Everyone's calling with ideas but I need someone to sit down and type and work it all out . . .

I know what I want, though. Modesty. A touching eagerness. Modesty with a touching eagerness with an assertion of experience—"She's been there—so she can help us here." Hey, that's good. Be a good tag line for radio spots, too. Modesty plus eagerness plus experience but also—also somehow a sort of feminine touch. Womanly. A little vulnerable around the edges, so I'm sympathetic, but tough enough to take it right in the face and keep on walking.

Where should I hold my big rallies?

Find out where Bobby Kennedy had his. Wait, Jack Kennedy had a great speech in New York—when he was running in '60, he had a great appearance in the Garment Center. The West Side, in the Thirties. I'd get big crowds—my people are the ones pushing the racks through the streets shouting "Watchyerback!" They're mine.

Speaking of watchyerback, what's Moynihan up to? I took on everyone he told me to hire; I've done nothing but be nice; but I don't see him much and he's out there embarrassing me supporting Bradley and saying Gore can't win. As if I cared about Gore, but appearances count . . .

Funny thing—everyone says Gore's too stiff, if he'd loosen up he'd win, blah blah. But I realize now he lost it the day Bill was impeached in the House and Al walked out onto the White House lawn with us for the victory rally and said Bill's one of our greatest presidents. We wanted him to say something like that of course, but—well, we never thought he'd go that far. I thought he'd be like Tipper and dummy up. Instead he looks like a lying lickspittle, like a weakling.

Doesn't matter what he does now, doesn't matter if he gets out there wearing a towel and doing the cha cha, he's over. Or almost over.

He hates me anyway, thinks I'm selfish. Well, welcome to the NFL, robot boy. The first rule of fight club is the rule of survival.

Anyway, less hope for Gore in New York means more hope for me; less money for him means more money for me.

Bradley. Bradley looks down on me. Bradley literally looks down on me, like he's gonna eat his lunch off a tray on my head. He makes jokes behind my back that he used to play for my favorite basketball team. He thinks I'm part of what he's running against. He thinks he's my moral superior.

Fine. Let him think he's better. He's got good numbers in New York, and if he gets the nomination we run on the same ballot and he gets my numbers up. Hell, Mr. Tall Story could win it for me in a close one. But he'll lose nationally while I win locally.

He'll go into retirement, I'll be on line for 2004.

But I have to concentrate on the here and now.

Got to think about the speech, the acceptance speech. It'll be major, full court press. I'll have reporters there from all over the world, more than Bill's had since his first inauguration. I'm going to take a leaf from Bill, and I'm also going to show him up. I won't go for the high eloquence, I'll talk more directly, throw in specifics, talk straight to the camera to get straight into the heart of that housewife from—from Massapequa.

Maybe that's where I should have a big rally—Massapequa. Suburban town in Nassau County, Long Island. Nickname: Matzoh-Pizza, but Mandy says don't call it that yourself, the old-timers say it but it'll be grating from an outsider. But keep it in mind—Jews, Italian, Irish—all the old I-countries. Third- and fourth-generation Americans, fruit of the European Diaspora.

God, my briefing books are good.

Famous people from Massapequa: Jerry Seinfeld, Ron Kovac, the Baldwin brothers—hey. Hey! I can get Alec and the boys to come with me. They love the Baldwins in Massapequa.

Wait. Joey Buttafuoco's from Massapequa. Bill's been compared to Joey Buttafuoco. Except, as some newspaper witch said, Joey has more class.

All right, put Massapequa aside for now.

The speech . . .

The carpetbagger issue . . . That's not so tough. It's the great thing I've got to knock down, but it's not so tough. Chuck Schumer said it last summer—"New York is full of immigrants who came here to follow a dream, and so has Hillary."

I'll look out at a bunch of recent immigrants and say something like, "All you've asked for is the right to pitch in and con-

tribute and make our state everything that it can be—and so have I. *You came here with humility, knowing that you wouldn't be accepted right away, but knowing that New York would give you a fair hearing—and so have I.*" Like that. *Then I say that we all came to New York, that immigrants have always come to New York, because it's the most open-hearted, good-spirited, and welcoming place in the nation. Well, something like that. Don't make other places look bad—they'll get mad at me in Arkansas. Hey—what am I worried about? I don't live there anymore.*

And then I should speak of me. Something like—who called in with this, I forget—"It's not where I'm from, but where I stand. It's not where I was born, but what I wish to contribute." *No,* "It's not where I was born, but what I believe." *I like that.*

Hey, what about a rally on Ellis Island?

No, too obvious.

Maybe somewhere in downtown Manhattan with the Statue of Liberty in the background. I actually look a little like the Statue of Liberty—why aren't the cartoonists picking up on that?

No, not Manhattan, they think all I care about is Manhattan anyway. Go to the outer boroughs, to Brooklyn or Queens, where the immigrants live. Festive and colorful and people of all races with all accents, all colors. We'll ask them to bring the flags of their old countries and wave them at the beginning, but we'll give them all little American flags, and when I announce they can wave them. That could be good.

Hey—I was at the unveiling of the plans for the new Eleanor Roosevelt statue. That could be a great place to speak—in the shadow of ER. "We are all in her shadow, but inspired by her legacy and spurred by her greatness . . ." Check it with Mandy. And Susan. And Harold. And Howard . . .

Maybe somewhere in the speech talk about the first time I came to New York, and saw the glittering towers, and fell in love with the energy on the streets, and hoped one day to return . . .

But—substance. Got to concentrate, got to get the substance right.

Got to show my state credentials. Something like, "I've been all across this state, from upstate Seneca Falls"—here do anecdote—"to the teeming streets of fill-in-the-blank, where I met a"—here do anecdote. ". . . And on my listening tour I heard you . . . I heard your dreams and your hopes, and you told me about your children. And I met a woman in fill-in-the-blank and she pointed at her daughter/son and said fill-in-the-blank . . . "

I have to remember to do what I did at Moynihan's farm— say I'm humbled and even amazed to be here, suggest I only got into this because New Yorkers drafted me. I got e-mails, faxes, calls—"Run, Hillary, run." It would have been ungracious to ignore these pleas . . .

But the other big thing: I have to speak like a romantic, like someone with the ol' Big Sweep. The vision thing. Recapture the old spirit of the sixties—evoke Bobby Kennedy. "And Bobby Kennedy, who came from another state and asked New York to consider him as its senator . . ." No, "who asked New Yorkers to consider him as their senator . . . "

Tie Bobby and the Sixties Vision to me. "The disappearing horizon of unfulfilled dreams, our never-ending quest for social justice . . ." Something like that. Camelot-y. What did Shrum write for Teddy—"The dream will never die." That kind of thing.

The point is I've got to mobilize the national fund-raising base and excite the national media. So my appeal can't only be local. I have to make the great liberal appeal, make it new again, resummon the old vision, the old feeling of excitement and possibility. At the same time, I can't say anything that jeopardizes the idea that I'm a moderate. It's tricky . . .

"To take the dreams of the past and turn them into practical wisdom . . ." What were those words Bill said about me—we were trying a line at the White House fund-raiser/birthday party last October. Bill called me "a genuine visionary, a practical doer." The crowd loved it.

That would be a great tag line for a commercial. Some big, deep, calm guy's voice: "Hillary Clinton. Genuine Visionary. Practical Doer."

But also it would be good for the speech—"Together, as genuine visionaries and practical doers . . . "

What else. Attack the Republican Congress in highly partisan terms and tie Rudy to it. Hit him hard—paint him as front man for rabid radical Republicanism. Paint him as pro-life. Paint him as pro-gun. Don't really blatantly lie, just insinuate, suggest, and hammer.

Susan keeps ragging me about giving up using government planes, she says I have to spring for a campaign jet. But I like the government plane and I've fought her all the way. But you know, a week or a month before I announce, we'll put out a big story about how I defied the Secret Service and insisted on taking . . . the shuttle. The Delta Shuttle. Or the US Air Shuttle. (Find out who's con-

tributing more.) I'll invite the press for my first day on the shuttle, we'll get a great picture of a stewardess tossing me a bagel, while I'm talking like a commuter to the lawyer squished in beside me. Or maybe the Metroliner—me, laptop up, writing and reading in coach. Note: Learn to use laptop . . .

What else? At least I got the look down; I'll stick with the dark suit, the open-collared shirt, the cute hairstyle, businesslike and pixie-ish at the same time. I never thought I'd get better-looking at fifty; I thought thirty was about as good as it was gonna get and it wasn't very good. But even I know I look kind of great now . . .

Got to keep staying in Chappaqua. What a mistake—I only want to be in the city, I wish I'd never bought it.

Got to try to get to know the neighbors, got to make friends with them so they'll start saying nice things to reporters. The guy down the block—when the van with my furniture arrived they started unloading, and some reporter asks the crowd, "What do you think is in all those boxes?" and the guy from down the block yells, "Evidence."

Then they start holding contests on the radio for what we'll call the house, and people start calling in with Mount Vernon Jordan, and the House of Bill's Repute, and Dogpatch on the Hudson, and DisGraceland . . .

Well, for now it will do. After the election I can unload it. Probably make some money. Sell it with a big ad—"Hillary slept here." It'll move. Hundred fifty grand profit at least. Better than cattle futures . . .

Settle down, concentrate. Stick in the here and now. Think about the acceptance speech . . .

CHAPTER EIGHT

I

Nassau County, on Long Island, is a fat, vote-rich county where Mrs. Clinton could do well. Bill Clinton carried it with 56 percent of the vote in 1996, compared to Bob Dole's 36 percent. Al D'Amato, on the other hand, almost won it when he ran for reelection to the Senate in 1998. He received 186,699 votes to Democrat Chuck Schumer's 188,335, losing the county by less than one percent. Mrs. Clinton will fight hard for Nassau County, and there are some issues there that will provide her with rich opportunities.

Every woman on Long Island worries about health costs, and every one of them worries, every day, about cancer. They think of cancer the way the pioneer women in John Ford movies viewed the Indian bands of the Old West—you're going about your life in peace, working and doing, and then, over there, on the ridge of the hill beyond—the shadow of men, in headdresses, on horses. And in the blink of an eye all is changed.

Long Island has a relatively high incidence of breast cancer, and Nassau County's is the second highest in the state: 118 cases per 100,000 population. For decades there's been a steady stream of news reports speculating on an environmental connection—pollution, pesticides, the water, the high levels of strontium-90 found in baby teeth collected on Long Island over the past few years.

Mrs. Clinton will address all this. There will be big meetings with people in pink ribbons, and some of it will be very emotional, with good people with real problems being comforted and encouraged by Mrs. Clinton. She will promise to be a voice in Washington for women and their special concerns; she'll back more funding for research, for environmental spending, for more studies. She will criticize unfeeling HMOs, and promise better health care. Her health care plan will probably be less a plan than an assertion: I care about your problems and understand them, and those Republicans in Congress, those unfeeling men in gray suits, those macho men with comb-overs, do not.

Crime could work as an issue for her, too. All of New York's suburbs are relieved about the plummet in crime locally, and so dramatically in the city, which they often visit and where many of them work. But they are anxious about upticks here and there—murder is up slightly in the city, and that tends to be a leading indicator of things to come. People are concerned about the prevalence of guns; they are concerned every day now when they send their kids off to school. So, though there is less violent crime now, it's not a settled issue; it still occupies a place in the American imagination. Mrs.

Clinton will hit hard and accuse the Republicans in Congress of being soft—of refusing to pass her husband's gun control measures, of balking at funding more police.

Will the women of Nassau County hear and respond? I say women because Mrs. Clinton sees them, of course, as part of her natural constituency, and because even though she does well with black women she has been disturbed to find out that she is not doing well with white women, who are not supporting her in the polls to the extent she'd hoped. Mrs. Clinton's ur-message to these women, her primary message, which will grow out of and be underscored by her appearances on cancer and crime, will be this: *Even though I come from another state, even though I've never lived here, even though I seem a product of the glamour and power of worlds far away, even though I have experienced dramas and circumstance in my life that seem frankly outsized and strange to you, and even though I'm the first lady . . . for all that, I'm a lot like you. I understand you, know your fears. And because I'm like you I'll know how to speak for you.*

Will women believe it?

II

In the heart of Nassau County, along the string of suburbs from Levittown to Babylon, is a town called Massapequa. Conductors on the Long Island Railroad call it out in a rolling bass—*Masssss PEE kwuh, Massssspeekwuh PARK*. Massapequa has been called "the new Peoria"—a town whose ethnic and demographic mix and whose particular voting patterns made it comparable to

Peoria, Illinois, in the 1960s—if it'll play there, it will play any-where.

Right now, in the spring of 2000, Massapequa is looking forward to swimming at Jones Beach and Gilgo.

I know because I go to Massapequa in the summer, to visit my friends. I grew up there. (There will be no surprise at the end of this chapter, I really did grow up there.) I will be there again this summer, on the beach at Tobay, visiting with close friends and barbecuing on the marina. And all around us there will be a bustling assortment of people—guys who look like bikers and men who look like middle managers, young fathers with three kids under the age of eight, grandfathers showing the kids how to catch crabs off the pilings on the dock. The women work in restaurants and own delicatessens, work in real estate offices and teach kids, are nurses and stay-at-home mothers, and most of them look like mothers and grandmothers because they are.

I think of the people I will sit on the beach with, my friends and their friends, and already I'm talking to them in my head. I'm talking to one of them anyway: Christine, who was my best friend when we were twelve and who is one of my best friends today, and who is thinking about voting for Mrs. Clinton but isn't sure. I'm imagining her and her husband, Bobby, and the kids at the beach on Saturday, August 26, 2000. I'm thinking:

Hey, missy, sitting on the beach and not going into the water because you don't feel you look so great in a bathing suit any-more and because—well, you're a mother, and you don't jump and splash in the waves anymore, the cold of the ocean feels icy now, and rough. Bobby still goes in, but this is just one of the many wonders of life: A guy hits forty-nine and he still splashes. He jumps into the waves like a smooth seal, comes up in a burst, turns and spits out surf.

The kids laugh, and you smile.

The noise of the beach, the radio, the crash and roar of the waves . . .

You're slathering SPF 40 on your pale freckled skin, and now Bobby is standing near you, arms crossed, looking at the waves and the kids and the lifeguards. Neither of you ever mentions this, but you both know what he's doing. He is standing watch, claiming and maintaining his hundred square feet of the beach, his blankets, his umbrella. He's making sure everyone's safe. He's looking to see if the lifeguards are paying attention, and if they're not he'll go and chat with them, tell a joke, ask a question, get them focused.

Three blankets, one for the kids and one for you and one for the coolers and whoever walks by. Your three girls are on the blue-checked cotton blanket gossiping and laughing and reading magazines. The coolers are full. You packed them this morning, Snapple and Rolling Rock and Diet Coke; you got up and made ham and cheese on poppyseed rolls, and turkey with mayo. Potato chips, Cheetos, Fritos, two boxes of SnackWell's. You packed more than you need because you don't know who'll come by, who'll stroll by and say hello, and you want to have enough. And they always come by because between you and Bobby and the kids and the school and your parents you feel like you know half the town.

It's 11 o'clock, you're all done with the morning packing and breakfast and getting out of the house and packing the car and getting here, and for the first time in five hours you can relax. Lunch won't be for an hour. You're sitting in the low green plastic chair and digging your toes into the sand. Bobby's talking with some kids near the water; the girls are rubbing on lotion.

And you pick up a magazine and she's on the cover. And you look at it. And you think what you always think when you see Hillary, which is that you don't really know what to think. If the piece you're about to read is highly critical, a real slam, you'll suspect it's exaggerated and partisan and mean. But if it's complimentary you'll think it's some puff piece, and you'll feel unsatisfied because you know, you can tell, there's something . . . not so great there.

And you don't think you're an important person but you are. Because what you're thinking as you look at the picture on the magazine at the beach at Tobay will determine the future of Clintonism in America.

You know she's trying to win you over, you can see it on the news, she's saying it all the time, "Our generation of women has shared the same experiences," and "I share your concerns."

And you wonder if it's true.

And I want to say, "Oh, kid, it is so not true. She is nothing like you."

And I know because I grew up with you, and because you—and please don't stop listening because you know a compliment's coming—you are something you never think you are, and that is: heroic.

You've made a good life. You have a lot to be proud of and you deserve to be proud and you are, but you don't make a big deal of it and in fact you barely notice it. You just tried to do your best, like everyone else. But somewhere you must know, on some level there must be a glimmering awareness, that with where you came from and where you got to, you've achieved something huge . . .

You're a teacher. Bobby was a teacher, too. But the money wasn't good and the kids were coming and he couldn't see a career path that was going to make it a lot better any time soon, so you kept teaching and he went to work at Con Edison. And you're doing good, you're doing fine. You own a house. There's a mortgage but it's a house, and you have two cars and enough food and you pay your bills on time. The kids will go to college. It will take help, scholarships and loans and grants, but you'll put all your savings toward it and take a second job and one way or another they'll get what they need.

And this, from the family you came from, is a triumph.

You are an American, Brooklyn born, and your parents joined the great migration from the city, the great migration of the hopeful European ethnics, the second- and third-generation Americans who left the city to get the first house in America, right here, on Long Island. Your parents, they were Sinatra generation—your father looked like Frank in *Some Came Running*, skinny and postwar. Your mother had a pixie haircut like Shirley MacLaine. Now they're in their seventies, but they're still rocking; they went to Atlantic City last weekend, saw a show and had dinner and she got into the slots.

That's what they are now, funny and shaky and still going. But then—back when they were twenty-five and thirty years old and they moved out to Long Island, where everything was square and flat and kind of lonely—back then they were overwhelmed. A house full of kids and low pay and a dicey employment history for Dad—your father's role model in those days was apparently Sinatra as Private Maggio in *From Here to Eternity*—and there were money problems and fights, depressions and dramas, and you kids barely made it through. But you did. You brought yourselves up. You went to McKenna Junior High, and then on to Massapequa High, and you had part-time jobs and rooted for the Lions and you got good marks and somehow, against the odds, you didn't drop out at seventeen and take a job in the city, you hung on and did your homework in a corner of the attic of your chaotic home.

And while you were there, Hillary Rodham was growing up in middle-class security in Park Ridge, Illinois, an affluent suburb of Chicago, and she had the things all kids want and many don't have—a highly functioning home, an orderly place where someone's in charge and someone makes dinner. Hillary at this point wasn't having anxiety attacks, the way you were. She was already used to telling people what to do. She was an old hand at running for office at school, a favored child marked for leadership, a Goldwater Girl—our families were for Johnson, we'd never even met a Goldwater Girl—with straight A's, a circle pin, and a ticket to the National Honor Society.

God, remember the Hillarys? The Hillarys would only be nice to us, would only look at us in the hall and say hello when they were running for senior council president. And then only because every vote counts. So she'd actually talk to people like us, and I wish I could say we told her to drop dead, but we didn't, did we? We were a little honored, because we knew what she knew: She was a superior person.

You just didn't think of yourself that way.

In 1969, right before college, you were working a job at the A&P, bagging and tagging, and America was going crazy with the war and the riots. And you were starting to see something.

You were developing a social conscience and you were starting to see the black kids from Amityville go off to Vietnam, and you knew something, because you had eyes. You knew that young kids with nothing, the kids from nowhere who didn't even have normal parents, didn't have the assumption that somebody would play ball with them or pay the rent—that they were the ones going off to fight for America. And the other, luckier kids who were white or had a father who was a dentist—they didn't have to go.

And it wasn't fair. And it wasn't right.

That year, 1969, was the year Bill Clinton of Oxford was dodging the draft. That's when he was writing his lying letters to the draft board, so he could get out of the war the black kids from Amityville, and the white kids from down the block, were fighting. Maybe one of them took his place. It's also the year Hillary, his soon-to-be-wife, was giving speeches at Wellesley— still an A student, still class president—and lecturing Edward Brooke, the only black man in the U.S. Senate, telling him that he was insensitive to the youth of our country. What we want, she said, is more ecstatic forms of living, more penetrating modes of being. You never heard of it at the time, that speech, but if you had you would have stopped putting the Wonder Bread in the brown paper bag for a second and thought, *Yeah? Maybe when you're not working at the A&P what you need is more ecstatic modes of being. I'd settle for a better job, or less work so I can study.*

Even then Hillary was on a higher plane—more abstract than you, more dazzled by big ideas. You were down-to-earth in your thoughts. It was easy for her, up in the thin air of high ideology, to shoot over from ideological right to ideological left. While your transit went from real-world bread and butter to . . . real-world bread and butter. Her sense of right and wrong became and stayed abstract, situational; your sense of right and wrong stayed grounded and concrete.

You got into a school, the State University of New York in Plattsburgh, and you got scholarships and grants and you worked as a waitress at night. And you did well, and met Bobby, and knew instantly. And when you got out of school you took courses toward a master's at night, at C. W. Post, and you got your first job teaching a sixth-grade class back home in 'Pequa.

And you and Bobby got married, and rented half of a two-family on Nanny Goat Hill, and saved. He was teaching at St. Pius, and working part-time at Holy Rood Cemetery in Westbury. He dug graves, and put down the sod on them when they were closed. Sometimes he'd bring you plastic flowers, and you'd laugh as you took them from his newly callused hands.

You learned how to teach. You were so happy. Nothing anybody asked you to do was too much, your enthusiasm and idealism made you forget any difficulties, and you looked at the other teachers with awe. Some of them had taught you when you were a kid. They were figures of respect.

And you did it for a few years, and then the girls came, and you stayed home with them. Then, in the mid-eighties, when the girls were in school, you went back. But already you could see something changing, and in time it got worse. When you were a kid people honored education in and of itself, and honored learning. But now you were seeing a new attitude—parents were starting to act like what their kids were there for wasn't an education but a degree—gotta get the kid's ticket punched.

And you saw something else. Every year during the eighties, every year they were loading more garbage onto teachers—more regulations, more programs, more rules about testing and tracking, more weird history, more p.c. stuff. The teachers had more burdens and responsibilities but less autonomy, less personal authority. It was as if the more burdens each level of government put on you, the lower the kids' scores dropped, which caused more new regulations to be applied.

So then the town decided they'd had it with the schools, and they voted to cut back the school budget. But then there were so many fixed costs and mandates on the schools that they had to start cutting back on the things kids really wanted and needed, the orchestra, the athletic program.

And the teachers' union seemed like the answer—it was growing in the eighties—but they were far away, and they didn't seem to know what's going on, either. And they wouldn't even allow the teachers—*their clients! the ones who pay the dues!*—to get the students to obey a dress code. They were showing up in cutoffs and undershirts. The girls in tenth grade, they were wearing little T-shirts—it was like everyone was in their underwear! But the union did one thing: They made sure teachers could dress bad too. So now they can wear jeans and T-shirts. A few years ago, during the contract negotiations, a teacher showed up wearing a T-shirt that had a little emblem that said "F*** Management." With the stars, but still. He wore it all day. No one said a thing.

So that's what you were doing, and learning, in the seventies and eighties.

That's when Hillary was exercising her first real power. She was in education, too. Her husband, the governor of Arkansas, had appointed her to head the effort to reform Arkansas's public schools. It was a slow-moving debacle, the kind that takes years to show it's a failure, and it was the very kind that was making your life miserable, because it was pushing and promulgating the kind of p.c. worldview that you were being forced to teach.

The worst part was Hillary's special baby—the governor's special schools for gifted students. They were all brought in from around the state to have a special summer semester together. And the point of it seemed to be to make religious students understand that their beliefs were backward and unsophisticated, that free market economies were bad and government control good, and that while ministers might be suspect and strange, left-wing radicals, feminists, and witches were not. One of the best parts, according to the notes one of the students took in 1980—the writer Joyce Milton saw them later—was part of a program that included a speech by Hillary herself, "who told student that she would trust big government over big business anytime." (If you'd heard anybody say that you'd say, "Yeah? I wouldn't trust either. But you force me to put my money down, I don't think I'd go with the bureaucrats.")

Another speaker, according to those notes, was "a physicist who said that science was the antithesis of religion, and no good scientist could be a religious believer. 'The book of Genesis should read, "In the beginning man created God,"' he told the class. Another speaker taught them that Christianity was anti-woman and anti-sexuality, and that the church was full of "fear and hatred of women." Even ten years later, in 1990, the school's speakers were assigning readings that called Christianity "a compost" and calling Christ's divinity "offensive."

But that isn't all Hillary was doing in the seventies and eighties. She was also making a killing at the Rose Law Firm. And making a lot of money sitting on boards—$15,000 a pop on the boards of Wal-Mart and TCBY, the yogurt company, among others. She was a highly desirable board member—an "education specialist," a lawyer, and the wife of a governor.

She was also making a killing in cattle futures in those days, investing a thousand dollars one day and reaping $99,000 in profits a few days later. She had friends who were doing business with the state of Arkansas helping to direct her trades. Later, when it became a scandal, she said she was just lucky. Then she said she figured it all out reading the *Wall Street Journal*.

And sometimes when she wasn't making money she was making speeches, as a public service. She often spoke against avaricious yuppies and the decade of greed.

She was flying high while you and Bobby were getting clobbered by inflation and taxes and child care. You were scouring *Newsday* for the ads from Giant and ShopRite, clipping coupons for the pork chops on sale. You didn't buy Tide and Cheerios unless they were on special, and going out to dinner was having buffalo chicken wings at happy hour at the local bar, with friends.

And then came the nineties.

And then '92.

And the Clintons were new to you, you'd never heard of

them but you saw them and they made a good impression, and you took a chance.

Why not? Bush seemed way out there in the ozone, like some sleepy old WASP who's detached from the world you live in. He seemed out of touch, goofy, and the economy was tanking.

And this Clinton, from nowhere—he was young and he seemed eager and hopeful and he reminded you of something.

He reminded you of 1960. He reminded you of when Jack Kennedy came down the Wantaugh Parkway in the motorcade on that cold October day, and suddenly your parents were interested in politics and you all went to the parkway and stood there and waved. Just to show him you were for him. Your father had a gray coat, and he watched as the motorcade went by, and he didn't know how to show he was excited so he just said, "Hey, hey!" into the air, and clapped as this brown-haired head went by in a limousine.

And you'll never forget it. And you realized, you absorbed for the first time, that you were part of something—you were part of America and politics and your parents were part of it, too, and it was—it was wonderful.

Bill Clinton made you feel like that, too. And you heard the stories on talk radio, "Slick Willy," the draft, the business scandal and the girls, and you thought, Okay, he's not perfect, they're not perfect, but they only made the kind of mistakes people make. And it's all in the past. Take a chance. Give him a chance.

And you did. And you know what they gave you. And it wasn't so great. Everyone's rich, that's good; he didn't do anything to mess up the economy; he kept Greenspan.

But the rest—the rest made you embarrassed in front of your kids. It made you feel like we should be embarrassed in front of the world. And in a funny way every time you see Clinton now you think of the teacher in the "F*** Management" T-shirt.

And you feel like maybe you had your last political fling, and you'll probably never have one again, and that's too bad but—well, maybe you're beyond flings now. And at least the economy's good. Turn the page.

But they won't let you turn the page.

Now she wants to be your senator.

Now she wants to speak for Massapequa.

Now she says she's just like you, her concerns are yours, she's a survivor like you.

But she is not like you. She never had to do the things you did, she never had to do it all uphill, and she is no feminist hero. The Eleanor thing you keep hearing—Eleanor was a nineties woman in the thirties, Hillary is a thirties woman in the nineties.

And that's it, really. The problem is not that she's a particular kind of candidate, but that she is a particular kind of person.

She grew up knowing something her parents had never known and you had never known: a completely secure life. But it didn't make her grateful—it made her presume.

The blood, sweat, tears, and toil of her parents created a world where she could find herself and her bliss. This wasn't everyone's world, of course. It wasn't the world of the razorbacks Bill Clinton left at home when he went off to Oxford. They left home, too, but they went to Khe Sanh, where they fought in a vain effort to keep other children from falling under the yoke of communism, where they would never have a nice day.

And it wasn't the world of young women whose parents had not been able to secure what Hillary had, and who were quitting high school and going to work at the fabric store at the shopping center down the highway.

And needless to say it wasn't your world.

But it was the world of the young Clintons, who were actually something new in history. They lived in a world that had never quite existed before—an aristocracy of the middle-class intellectuals, peopled by the heirs of genetic and financial privilege who declared themselves king and queen of the future. Freed from humdrum concerns about getting killed or not having enough money or enough intellectual or emotional resources to get to college, freed from the worries that held back their parents, the grunts in the battle for affluence, freed from all that *they forgot to be grateful to the place that had made them, and that ensured their rise. They forgot to love it.*

They lacked the wisdom to see their luck as something to be humble about. Instead, it mutated within them into a sense of ongoing entitlement and superiority. They were going to educate people out of their old-fashioned, backward, racist, sexist, hopeless ways. That's what Hillary really meant when she said she wasn't some little Tammy Wynette. She meant: I'm not some ignorant big-hair girl working the counter at the Piggly Wiggly; I went to Yale Law.

They were lucky. And a funny thing about the long-term lucky is that they often come to think not that they were blessed but that they deserve it. That they lived right, while other people, unlucky people, did not. They think the fact that their lives are good, that they are attractive and articulate and have prospects means that they deserve these things because they're—well, better. The proof is in the pudding.

And when you think like that it hardens you. You can talk on and on about compassion and tolerance, but it hardens you in a way that your parents' experience—which was actually hard—didn't quite harden them. You look down on the girls with big hair and the boys with big trucks. You start to understand that the world would be a little better ordered if you could tell other people how to organize their lives.

And that is the one thing that people who know Hillary *always* say about her: that she thinks people have to be led and guided. Implicit, always unstated but always understood, is that they should be led and guided by: her. And her friends.

And so Hillary Rodham and Bill Clinton had their new world, an abstract and marvelous world of identity politics and village bureaucrats and government calling the tune—and those governments, at whatever level, county, state, or federal, would be dominated by the other middle-class intellectuals, the networking revolutionaries, the radicals with Rolodexes.

Are they arrogant and opportunistic? Yes, they always were, but the headline on the Clintons is that they never grew out of it.

They never graduated from the sixties.

And she never graduated from herself.

To say the problem with her candidacy, then, is the carpetbagger issue, is to miss the point, and to be unkind to carpets. Carpets are things that get walked on. But her boots are made for walking, and they'll walk all over you.

You wonder if she is like you. But she is not.

You were a boomer and she was a boomer and you faced the

same choices—but you went different ways. You didn't have a lot of options, but you tried to open every door; she had every option, but didn't walk through those doors herself. You found a good man and did it with him, she attached herself to a charismatic character and did it through him. You wanted a decent life, she wanted power. You made a marriage, she made a deal. You became a citizen, she became an operator; you became someone who contributes, she became someone who connives to tell the contributors what to do and how to do it. She portrays herself as a victim, but she's a victimizer. She says she is a survivor, but she's one of the people you had to survive.

She has neither your class nor your courage. If you want a woman to represent your point of view, then you can find one in New York, which has plenty. Don't fall for the one who only wants to use this place as a stirrup to climb her way onto a horse called the presidency.

She doesn't know your concerns, and she doesn't share them, either. She is not like you. She was never like you.

CHAPTER NINE

HILLARY CLINTON OFTEN SPEAKS ABOUT CHIL-dren's issues. She has written quite a bit about children, and her claim to expertise is based primarily on what she has written. Interestingly enough, though, most of what she has written is not up there on her website and never quoted in her columns. And the most striking thing about her backlog of articles and essays is that she has never repudiated them, never announced that she's changed her mind about her positions.

This is surprising, because they are quite radical.

Mrs. Clinton's writings reveal a particular mind-set, particular impulses, and particular assumptions that may suggest a great deal about her future political actions and decisions. (Her writings certainly signal her intentions: She was calling for grass-roots political pressure for the creation of a new national health care system back in 1972.)

As David Brock wrote in *The Seduction of Hillary Clinton*,

the general theme "running through [her] writings is the desirability of transforming social relations and politicizing family life through bureaucratic and legal processes, and enhancing the paternalistic role of state agencies and trained experts in all spheres, especially those connected with children. The tone and substance of her writings are those not of a Marxist revolutionary but of a Swedish-style technocratic socialist."

The first article Hillary wrote on children's issues, "Children Under the Law," was published in 1973 in the *Harvard Education Review*. After announcing that she would like to be "a voice for all children," Hillary asserts in the piece that we must change the way children are viewed and treated under the law. They should be viewed as "child citizens." They should no longer be treated, as American legal tradition has long held, as not competent in legal proceeding by virtue of their age. They should instead have all the procedural rights guaranteed in the American legal system to adults. The old legal presumption that there is an identity of interest between parents and children, she contends, should be rethought. She praises Supreme Court Justice William O. Douglas's dissent in a Supreme Court case that ruled that parents of Amish children have a religious right not to send their children to high school, but to home-school them. Douglas disagreed, arguing that children's views on questions such as this should outweigh those of the parents, and that if Amish children in particular would like to become artists or astronauts, for instance, they would have to break from the Amish tradition; if they are "harnessed to the Amish way of life" their lives would be "stunted and deformed." (That must have been news to the Amish, who think of themselves as parents who bring their children up well, and who in any case have the right to raise their children according to the tenets of their faith.) Hillary seconded Douglas's dissent, saying that children should be able to make independent, "responsible decisions on matters of religion and education."

More tellingly, perhaps, Hillary advocates liberating our child citizens from the "empire of the father"—reflecting the anti-patriarchy claims of a particular kind of feminism. She argues further that children are locked into an unjust dependency relationship within their families, and need the assistance of the state in protecting their interests.

"The basic rationale for depriving people of their rights in a dependency relationship," she writes, "is that certain individuals are incapable of or undeserving of the right to take care of themselves and consequently need social institutions specifically designed to safeguard their position ... Along with the family, past and present examples of such arrangements include marriage, slavery and the Indian reservation system."

Now, those are famous words, and have been much spoofed and much written about. But when you look at them you have to wonder if she means it—if she meant it then and means it now. We have all thought and written things that twenty years later make us roll our eyes. But again, Mrs. Clinton has never disavowed these words. When pressed on them, in the 1992 campaign, she simply insisted that Republicans were viewing them through a partisan lens. Which is a charge, not an explanation.

In the article, she admits to a political dynamic in her discussions of children. She writes that the children's rights movement "highlights the political nature of questions about children's status . . . The pretense that children's issues are somehow above or beyond politics endures and is reinforced by the belief that families are private, non-political units whose interests subsume those of children."

She betrays her view that law is not a neutral set of principles but a social construct subject to many interpretations; she argues that the position assigned to children under the law should be seen as "part of the organization and ideology of the political system itself."

Who should decide when state intervention is right and necessary in the event that a child should need to be removed from his or her family? Here Hillary puts forth an unusual solution. She charges that state agencies and the courts often rely on "middle-class values" to judge a family's child-rearing practices. This, she says, is unfortunate. Instead, decisions to intervene should be entrusted to "boards composed of citizens representing identifiable constituencies—racial, religious, ethnic, geographical," which would decide when parental rights over a child should be terminated.

Now, many things might be said about this, but one of the most intelligent came from the writer Christopher Lasch, generally regarded as a liberal on public policy issues. He wrote, in an analysis of Hillary's views in *Harper's*, that while they might look attractive at first, "a careful reading of [her] argument . . . shows that she objects to the family much more than she objects to the state . . . Although she warns that the state's authority must be 'exercised only in warranted cases,' her writings leave the unmistakable impression that it is the family that holds children back and the state that sets them free. Her position amounts to *a defense of bureaucracy disguised as a defense of individual autonomy.*" (Italics mine.)

In 1977, Hillary wrote an essay called "Children's Rights: A Legal Perspective," which was included in a collection of essays on children's welfare. Here she acknowledges that some experts warn that the move to declare children to be bearers of the rights given to adults would give rise to a new class of rights that would be difficult to limit. But she says she does not share this concern. She asserts that the decision of a minor "about motherhood and abortion, schooling, cosmetic surgery, treatment of venereal disease, or employment and others where the decision or lack of one will significantly affect the child's future should not be made unilaterally by parents. Children should have the right to be permitted to decide their own future if they are competent." She goes on to suggest that children might assert their rights to "grow up in a world at peace" under a United Nations declaration, and says that "children and adults might have special standing to question the proliferation of nuclear power or junk food . . . "

Again Lasch's response is instructive. He points out that when children are granted the legal right to speak fully for themselves, it opens the possibility "for ventriloquists to speak through them and thus to disguise their own objectives as the child's." Who would be the ventriloquists? People like Hillary, filling the ranks of government bureaucracies. As he points out, for Hillary the only reliable index of progress is a proliferation of state-managed and taxpayer-funded programs.

A year later, in 1978, Hillary wrote for *Public Welfare* of what she called the "myths" that hold back "the development of a realistic family policy in this country." Among those myths is "The myth of the housewife whose life centers only on her home [which] is effectively dispelled by statistics demonstrating that the average school child now has a working mother."

One is struck, when reading Hillary, by how often she mis-states an argument in order to refute it. Who believes now, or believed in 1978, in a myth of the housewife whose life centers only on her home? What about the reality of the mother whose primary concerns revolve around her family, children, husband, home, neighborhood, and children's schools? There are many such women, and good thing, too, because as much as anyone and certainly no less, they keep America going.

Hillary Clinton has in the past few years been careful to insist that whatever path a woman chooses, what is important is that she be able to choose freely. But from the beginning of her career, she has betrayed a certain condescension toward those who stay home with their children and exercise their right to leave, not enter, or delay entry into the workforce. One senses she sees them as timid, as if they have betrayed in their decision a cowardly response to feminist imperatives, and have somehow failed the collective.

In 1977's "Children's Policies: Abandonment and Neglect" in the *Yale Law Journal*, Hillary criticizes those political leaders who have failed to involve government more directly in family policy. She recommends the establishment of a new federal agency for children's needs. Politicians, she explains, are too busy thinking about "missiles" and the national defense to understand the need for "a federal policy for children." She criti-cizes those who argue for caution, and who contend that the burden of proof in the argument over more federal involvement in the lives of children should be on those who say that parents are not competent to raise their children. Hillary says this is just an excuse for a lack of daring. "Legislators and executives take risks of all kinds when they decide to build a nuclear plant or introduce a deadly pesticide or advocate no-fault insurance." If legislators really cared about children they'd take a risk and develop a new national policy for them.

But to get a real sense of Hillary's thoughts on children, and how the legal realities of childhood should be treated in America, the best source is Joyce Milton's fact-filled biography, *The First Partner*. Milton had written a laudatory children's book about Hillary, but found in time that the more she watched Mrs. Clinton and looked into her life, the more reservations she had.

By 1994, the health care failure, the travel office scandal, Waco, chatter about "the politics of meaning," and "posing for glam shots for *Vogue* magazine" had put Milton off. She decided to write a full-scale biography, and the book she produced is as much an indictment as a biography—and highly readable to boot.

In 1977, Hillary was hired by the Carnegie Council on Children, a blue-ribbon panel of experts assembled in part to respond to the work of the respected sociologist Uri Bronfenbrenner, who had recently compared child rearing in the Soviet Union and the United States and pronounced the Soviet system superior.

As Milton relates, the Carnegie Council report, and the sections that Hillary researched and wrote, are "must reading for anyone who seeks to understand Hillary Rodham's vision of the future of American families."

The members of the panel began with the assumption, as Milton puts it, that "the triumph of 'the universal entitlement state' was inevitable, and the best thing Americans could do for their children was hasten its arrival." Government would take responsibility for more areas of our children's lives than merely their schools; as the American nuclear family declined, divorce would not be a concern, because government workers will provide support whether a family is intact or not.

The report, which was entitled "All Our Children," called for everything from universal health care to full employment, if necessary through government-created jobs. The cost and consequences of these programs were hardly addressed.

As Milton notes, "All Our Children" offers "a blueprint for undermining the authority of parents whose values the authors consider outmoded."

Hillary worked on the chapter entitled "Protection of Children's Rights," which states that "it has become necessary for society to make some piecemeal accommodations to prevent parents from denying children certain privileges that society wants them to have." The piece goes on to call for laws allowing children to see doctors for pregnancy and drug use without parental notification, and preventing schools from "unilaterally" suspending or expelling disruptive students.

But that is just the beginning.

The panel also called for the creation of a new class of "public advocates." "In a simpler world," it announced, "parents were the only advocates for children. This is no longer true. In a complex society where invisible decision makers affect children's lives profoundly, both children and parents need canny advocates. What if all parents made relatively small financial contributions to such a cause? It would provide a politically insulated fund for lawyers, ombudsmen, agency monitors, and even attempts at legislative reform."

The report suggests that professional "child ombudsmen" be placed in public institutions to enable individual children to hire "decently paid" private attorneys to represent their interests. What would such attorneys do? Well, the report suggests, they could bring class action suits to hold corporations liable for future damages their businesses might cause to today's children. As for the ombudsmen, they might consider "how to make the rewards in our society more just and how to limit the risks of technology . . ." They might also "ask about tax reform, about reorganizing health care, about racism, about sexism, about energy—all for the sake of our children."

"All for the sake of our children"? It sounds more like all for the sake of attorneys like Hillary and her friends, and all for the sake of left-liberal political action.

As Milton writes, "It is hard to find in this much evidence of great concern for actual, living children, including the 'relatively small number' who are victims of neglect and abuse. Rather, this is the voice of people who think they know all the answers, and want to use children as a tool to impose their own will on others."

Even in *It Takes a Village*, published during Hillary's tenure as first lady, Mrs. Clinton's policies for children proved unusual enough to draw some fire from professional analysts of government policy such as the respected social critic Jean Bethke

Elshtain. In a 1996 *New Republic* essay prompted by Hillary's book, Elshtain wrote that Mrs. Clinton's policies invite "recourse to the courts and more generally . . . [invite] moralistic overreach, paternalism, and a limitless extension of sympathy that casts a pall over political debate because it transforms one's opponents into nasty depredators who mean to do children harm."

Milton sees Hillary's policies as betraying the mind-set of a left-wing political operative who is using children to create a new liberal order of government-mandated involvement in what might be called the private life of America. And David Brock, in his biography, sees an arrogance in her proposals for children—an assumption that Hillary and her friends know what's good for us and, once in power, will give it to us whether we want it or not. Like a Swedish social planner, she knows the way, and believes she has the right to take the resources of the American people—and a portion of their personal autonomy, too—to get us to her vision of a better world.

A final thought. For those of us who oppose Hillary, one of the things we've always been most impressed by, that we give her a lot of credit for, is her daughter, Chelsea. That young lady has always seemed—including the two times I met her, when she was a child—a nice, kind, and generous person. Such qualities in a child (or anyone else) are an achievement.

And the funny thing is, Mrs. Clinton raised a nice, smart girl who wants to be a doctor . . . and did this *without the interference of the state.* Chelsea Clinton was sent to private school—not a government-run public school but a tony and expensive private school—in Washington, D.C. She did well and made friends, graduated and went on to Stanford. Chelsea was not raised with the help of child advocates, child ombudsmen, or government-appointed attorneys. Chelsea was also allowed to function, according to all reports, as a child; she was not treated by her parents as a "child citizen." They took an active part in decision making in her life, rather than ceding such authority.

Mrs. Clinton, apparently, had no interest in extending her policy prescriptions to her own child, even though she seems convinced that they should apply to other children—such as, perhaps, yours. One can't help but wonder why she thinks that what is good for her might not be good, also, for others.

Does Hillary think that most people are not as capable as she is? Does she think that most mothers and fathers do not deserve to have the autonomy and authority that she and her husband, as parents, have enjoyed? Does she think other parents are less able to make parental decisions than she and Bill were?

How odd that she would think this way, if she does.

But, you know, you can wind up thinking a lot of odd things about other people when you have spent your entire adult life, since the age of seventeen, in Ivy League colleges and Ivy League law schools and government cars with government drivers who take you to your taxpayer-funded mansion where you don't bump into too many rank-and-file normal people. And if on top of that you spend most of your free time with other people who went to elite schools and law schools—and who themselves are detached from normal American life—you can begin to think that other people, the ones outside your peer group and social circle, are lesser people—less educated and intelligent, less civilized, certainly less successful and celebrated and affluent. And you can start to assume odd things about these people you don't know. That they probably spend a heck of a lot of time beating their kids and abusing them and shutting them up in Amish farmhouses and evading the law, which only wants to save them . . .

But more to the point, for the purposes of this book: It is fascinating that Hillary Clinton has never gone to any lengths to disavow the things she has written and argued.

It is also intriguing that her entire reputation as an expert on children's issues—or, as she puts it on her website, as "a lifelong advocate for children and a champion of their rights"— rests on her writings about children quoted here, and on her involvement in the Carnegie panel for children. So when she calls herself an expert, this is what she's talking about.

One last thing: She stopped writing the kind of things I've quoted in the early 1980s, and has not talked about them since 1978, the year her husband first ran for and won the governorship of Arkansas, and began their long climb to the White House. She doesn't deny or disavow or second-guess them, she just doesn't talk about them. As if she doesn't want anyone to know.

CHAPTER TEN

THE WAYS AND HABITS AND TEMPERAMENT OF a politician have a serious bearing on what he or she does in office. And one of the things that has struck many observers of Mrs. Clinton is that she often displays in her political life what appears to be a startling lack of wisdom—the kind of wisdom that has to do with generosity and patience and the knowledge that in the long run you have to be thinking about . . . the long run.

She is famous for her toughness. According to all the White House memoirs and looking-back interviews, it is she who always stands and says: Don't turn over the documents, don't admit anything, don't settle the lawsuit, don't go for a special prosecutor, don't give an inch, hire an investigator. Whoever makes a charge against us will be designated a target. And what do we do with targets? We bomb them.

*   *   *

There is good toughness and bad toughness. Hillary's good toughness has brought her to the White House and allows her to function there each day in spite of the pressures that are endured even by first ladies who are not followed by constant charges of personal and political corruption. Hillary has the kind of toughness that lets her wake up each morning no matter what, drag herself out, and get in the game. This is good tough. Everyone needs it, and she has it in abundance.

But she has bad tough in abundance, too. The kind of tough that is so much a habit, a reflexive reaction, that so quickly assumes bad motives, that so readily sees dishonesty and hatred and jealousy, and that is so quick to fight, level, and kill that it isn't tough anymore, it is hard. And when you get hard you do hard things, and those things have a further hardening quality, and you become harder still.

A paradigmatic Hillary moment, as recounted by George Stephanopoulos in *All Too Human*:

It is 1994. The subject is Whitewater. The *Washington Post* has been asking questions about the Arkansas land deal. Hillary has convinced the president to "stonewall," in Stephanopoulos's words. The strategy has resulted in failure. The scandal continues; now the Congress wants documents, and there is talk of a special prosecutor.

"All day Tuesday, we had a rolling Whitewater meeting in Mack [McLarty]'s office," Stephanopoulos writes. George, who had "delivered the party line" against appointing a prosecutor on *This Week with David Brinkley*, was arguing now, he says, that the White House should request a special counsel. The top staff reaches a consensus: If we don't ask for a counsel now we'll have one forced on us.

A White House lawyer is dispatched to Hillary to discuss the matter. She "shut him down." Harold Ickes and McLarty go to her. "The answer was still no."

The staff gathers in McLarty's office to review the situation. Hillary walks in. "I think this is a meeting I ought to be at," she says, giving everyone a look. She says not to give Congress the documents they're demanding. "If we were as tough as the Republicans, we'd band together and beat them back." More talk, then her directive. "I don't want to hear anything more," Hillary says. "I want us to fight. I want a campaign now."

If we were tough . . . beat them back . . . I want a campaign . . . I want us to fight . . .

And so a scandal became a political catastrophe that to this day isn't over.

Another moment, this one recounted in Gail Sheehy's book on Mrs. Clinton. Bill Clinton, in the midst of the Monica scandal, is making his disastrous August speech to the nation. Hillary watches upstairs in the White House with several aides. She had urged her husband to "come out and hammer Ken Starr."

When the president finishes, Mrs. Clinton "jumped to her feet and applauded the screen. 'Yes!' she said. 'That'll show them.'"

That'll show them.

Another Hillary scene, this one from Jeffrey Toobin's *A Vast Conspiracy*.

Hillary has just given her furious interview to Matt Lauer on the *Today* show, the one in which she charged the president's troubles were due to a vast right-wing conspiracy. She has returned to the White House, where her friend Linda Bloodworth-Thomason congratulates her on her television appearance. Hillary shoots her a look. "That'll teach them to fuck with us," she says.

These are her behind-the-scenes leadership characteristics—an almost antic belligerence, a pugnacity that allows her to countenance low deeds because the world is full of low people who want to get the Clintons. Smear someone in the press? Sure. Lie to investigators? Say you don't recall. Lie to the press? In a shot. Lie to a grand jury? Say you don't remember. Ruin a reputation? They deserve it. Turn over a document? Stonewall.

And it is this part of her nature, the unchecked belligerence, that allows her to indulge what appears to be another part of her nature—magical thinking.

Magical thinking is believing that a problem will be solved not by serious, realistic thought and sound effort but by—by whatever. By luck, by the continuance of your own good fortune, by a shaft of light, by a roll of the dice, by the magical finger of God touching your magical destiny.

When Hillary told Bill not to settle with Paula Jones she was indulging in magical thinking. Maybe Paula would fold, she must have thought; maybe Bill didn't do it; maybe some guy would come forward and reveal that Paula Jones had once attempted to entrap and sue him, too; maybe Paula will tire; maybe her supporters will go away. Maybe—maybe our luck will hold. The Rose Law Firm billing records existed, they were real and palpable things you could hold in your hand. Let's hide, or rather misplace them. Maybe they will never turn up. Maybe no one will notice them. Maybe if they're found my fingerprints won't be on them. Maybe Ken Starr will disappear.

Maybe when Web Hubbell was taped saying, "I'll just have to roll over one more time," people will think he's talking about having trouble sleeping in jail. Maybe no one will notice that I made $100,000 almost overnight in cattle futures, and maybe if they find out they'll believe I learned how to invest at my father's knee.

If childhood is inescapably the forge of the future adult, perhaps one key to Hillary Clinton's character can be found in a mythic childhood story she has shared with biographers and profile writers. One day when she was in grade school, she recalls, she came home crying to her mother. There was a little girl who kept beating her up every time she went out. Hillary wanted comfort. Instead she got a directive, of a particular sort. "There's no room in this house for cowards," her mother informed her. "The next time she hits you, I want you to hit her back." Hillary, as she reports it herself, did just that. Thereafter the neighborhood boys let her play with them.

If this story is true, it's pretty clear what Hillary learned. To play with the tough boys, you gotta play tough.

But it's tempting to wonder what American politics in the nineties would have been like if her mother had said "Aw, hon, come on in and have some hot chocolate," or "Well, the Lord tells us to turn the other cheek," or "Well, they're acting immature, and you might someday have to point that out. In the meantime I'll ask Daddy to teach you how to box so if you absolutely have to you can deck the little jerk—but only if you have to."

Or even what Ronald Reagan told his daughter Patti. When Patti was in grade school there was a bully on the school bus who picked on her every day, pushing and saying mean things. Patti went to her father. He told her that the bully only wanted her attention, so the best thing was to try not to give it to him. He showed Patti what to do. "Try and get my attention by annoying me," he told her. She did. She talked loudly. He stared ahead with a pleasant face. She waved her arms and made faces. He pretended he was looking out the window at nice scenery.

Patti remembers this for two reasons. One is that it scared her—decades later she saw it as a metaphor for her relationship with her father, her struggle to get his attention and her sense that he was otherwise engaged. And the other reason she remembers it is: It worked. She got on the bus, ignored the bully, and in time he got bored and redirected his attentions.

But Hillary's mother told her to be tough, and Hillary is. And the key word in the story is *coward*. "This family has no use for cowards." Is that what a little girl who doesn't want to punch somebody in the face is? Is that what Hillary came to think a person who won't hurt, or smear, or destroy another person is?

I don't mean to make too much of this, but think of how different history might have been if Hillary had been told that day to be patient, count to ten, come color this book with me. *Bill, you were probably no gentleman to Paula—apologize and get this behind us. Bill, later I'll hit you over the head with a frying pan, but for now, tell the truth about that intern, we can't put the country through months of trauma and embarrassment. Bill, I'm handing over the billing papers, and there's something I didn't tell you about Whitewater. Bill, people are walking in here with $50,000 checks and it's against the law and I want it to stop. Bill, I got up for a glass of water last night and there were complete strangers in the Lincoln Bedroom, and if this is how we're raising money then we're doing it wrong. Bill . . .*

CHAPTER ELEVEN

Another word on campaign coverage. It is interesting that candidate Hillary Clinton has a busy campaign schedule and speaks in public almost daily, and yet there are very few press reports that contain extended quotes of her remarks.

This is odd; Mrs. Clinton's is such a striking candidacy, and is so heavily covered, that reporters would be naturally inclined to cover her remarks at some length to make clear her views, give a sense of the flavor of Hillary on the stump, and add to the historical record.

But with Mrs. Clinton, reporters seem to limit the number of quotes they use. They tend to concentrate instead on the venue of the speech, the atmospherics surrounding it, and pictures of the event.

There's a reason for this.

Almost a year into her candidacy I gathered as many Hillary speech quotes as I could, and reviewed them. They are relentlessly banal. Here is a fair sampling: "Governments must put children first"; "Every time we pay tribute to art, particularly to art in a public place, we know it will cause a lot of thoughts to be thought and words to be spoken and ideas to be sparked"; "Each of these precious children is a child of tremendous potential, potential that can be unlocked in the first years of life, or locked away for a lifetime." From her October 1999 visit to a Jewish school in Warsaw, Poland: "This school is very important to you but also for other reasons—your songs of peace show how hard you have worked to make your voices heard in today's Poland"; "The Jewish traditions which have enriched Poland for centuries will again enrich Poland in the next century." On the Italian leg of the same trip: "I see the desire of so many to try and assert their identity against the mass culture fueled by globalization"; "We must bring a new cultural literacy and cultural respect to all our development strategies"; "We must remind children of their heritage and teach them to respect culture."

When you read these quotes, you wonder what it's like to listen to them for thirty and forty minutes at a clip; you can almost feel reporters straining to find something, anything, that will brighten their copy and show their editor they're not talentless hacks who don't know an interesting comment from a cliché.

Why would a dramatic candidate for high public office in a media-drenched state like New York make it a point, every day, to say nothing that would be the least bit memorable or interesting to potential voters?

Because she has a strategy. The strategy is to use words not to pierce through the fog of voices and sounds in which we all live, but to add to the fog. She is using words not to reveal but to conceal, not to make clear but to confuse. She is not trying to communicate her thoughts, ideas, and plans; she is merely trying to communicate an impression with pictures.

For as long as she can in the campaign, she will limit her communication to the symbolic. She will continue to visit schools and sick people and town squares, and she will deliberately say nothing interesting during these visits. She will let cameras take pictures. Later that day, on the television news, the local anchorman will compensate for the fact that she didn't say anything by providing the kind of narrative journalists in such circumstances feel forced to provide: "Hillary Clinton visited a Harlem day care center today and said she 'cares deeply about our children.' Calling them 'the future of our nation,' she met with students"—here there will be a happy shot of Hillary laughing and hugging a pigtailed seven-year-old—"and gave the headmaster a much needed surprise, a $2 million federal education grant recently earmarked for inner-city schools."

Mrs. Clinton does not want words to compete with the videotape. She wants only the videotape, because it communicates what she wants communicated: Hillary is a warm woman who loves children, and not the ruthless operator she's painted as by people who aren't warm and don't care about children.

There is another reason she says nothing interesting in public. It is that if she declares, or allows herself to be drawn out about, her real philosophy and views, she enters a danger area. Her political impulses are left-liberal, and if she declares this she will lose votes. But if she denies it, she endangers the farthest left corner of her base. In a close race, those highly motivated troops could make the difference. She will continue to say nothing because she wants not to explain her stands but to obscure them.

This is not an example of the collusion of the press but the desperation of the press. Local TV stations free up a reporter and a camera crew, and budget them, to accompany Mrs. Clinton on her trip to the day care center. They have to get a story out of it or they've wasted their resources; they also need tonight's Hillary story. To get the story they have to provide the narrative that makes up for the empty spot created by what Hillary didn't say.

This will probably work very well in the campaign, and for a long time. And her prospective opponent, Rudy Giuliani, may well make it even more effective by providing too vivid a contrast: he'll tell everyone who asks every opinion he has. *("Of course the Brooklyn Museum should stop showing the Blessed Mother surrounded by animal dung! Well, if you don't think so, maybe you're a bigot yourself. Go home and take your meds.")* He is blunt and relatively candid, but may wind up looking, in contrast to Mrs. Clinton, wild and shrill. She is misleading and manipulative, but may look laughing and warm.

All of this reflects the Clintons' longtime media m.o.; it is their own variation, or twist, on what they learned from watching Ronald Reagan in the 1980s. Reagan spoke directly and consistently for thirty years about his philosophy and plans and positions. He used words not to create a fog but to cut through it, to make clear his stand on communism, on taxes, on foreign affairs and domestic issues. Everyone knew where he stood, and could back him or oppose him based on that knowledge.

People like Mike Deaver, who worked for Reagan, saw to it that Reagan's speeches and appearances were well choreographed—crowds, cheers, balloons, and bunting. This was state-craft as stagecraft, and was usually effective. (There's a reason he didn't say, "Mr. Gorbachev, tear down this wall," in a speech from the Oval Office. He said it, of course, in front of the thick, ugly Berlin Wall. The picture underscored the meaning of the words. For the same reason, John Kennedy said he was a Berliner in front of a crowd massed in the center of Berlin, not in the State Department auditorium.)

Using pictures to communicate where you stand and why you stand there is honorable, and a necessary acknowledgment of the reality of mass media. But Mrs. Clinton uses stagecraft to hide and obscure philosophy and positions, which is not honorable at all.

CHAPTER TWELVE

OPEN BEFORE ME IS MY SON'S COPY OF *PROFILES in Courage*, the old hardcover edition from Harper & Brothers, 1956, that I picked up at a book fair many years ago. It seems to me tender-looking now, its cover torn and dry, the red, white, and blue lettering faded. On the back is a portrait of the author, the young senator John F. Kennedy of Massachusetts, seated in his office, a properly professorial pair of reading glasses folded in his hands.

I turn to the preface and hear the tenor and tone of the old America. "Since first reading—long before I entered the Senate— an account of John Quincy Adams and his struggle with the Federalist party, I have been interested in the problems of political courage..." The chapter headings, too, seem elevated. Daniel Webster's begins with "...not as a Massachusetts man but an American," which is how he fought, uphill and only for a moment successfully, to hold the Union together. "The magistrate is the servant not . . . of the people but of his God," heads the section on John Quincy Adams, and his struggles with tariffs, and war.

I keep this old book on a bookcase above my computer, near some others that seem, to me, to go with it—a few memoirs, *The Education of Henry Adams*, a little book of children's poems by Rosemary and Stephen Vincent Benét, *A Book of Americans*. There's one about Abe Lincoln's mother—*"If Nancy Hanks / Came back as a ghost, / Seeking news / Of what she loved most, / She'd ask first / 'Where's my son? / What's happened to Abe? / What's he done?' . . . "* And one about Dolley Madison: *"When the British began / To cut more capers / And burned the White House / She didn't take vapors. / The roofs fell in / And the cut glass burst— / But she saved George Washington's / Portrait first . . . "*

These old books seem to me somehow fragile, and I feel protective of them, as if they represent something old and not to be lost, something we don't want to lose—a sense that men, frail, mud-made men, and women can produce great triumphs when they first triumph over themselves. Most of Kennedy's subjects managed to defeat within themselves, if only for a moment, what most of us have and could do without, and what the political personality has in arguably greater abundance—the desire for respect, for admiration, for what Thomas Hart Benton called "the bubble, popularity." In defeating those desires, in turning away from the feeding and nurturing of them, in, in fact, sacrificing them, each of these men produced an unforgettable triumph, and did something authentically great. We read these stories, and poems, because they inspire. We think: *Sam Houston was vain, intemperate, and loved fame, and yet he was willing to lose everything, and did, to keep the Union whole. Maybe I, with my own big flaws, can do some great thing.*

What do the Clintons do? Do they inspire us? Or are they rather a cautionary tale about what you can become when you cannot sacrifice your needs, or quell your hungers, and never quite manage to summon the grace to put principle or country before self?

On the wall beyond my desk is a picture from a supermarket tabloid. It was taken about five years ago, and caused much comment when it was published. It is a picture of two close friends of Bill and Hillary Clinton's, Linda Bloodworth-Thomason and the actress Markie Post. They are holding hands and jumping up and down at where fate had put them. It had put them in the Lincoln Bedroom. They were jumping up and down on Lincoln's bed.

It seemed to me to symbolize the Clinton White House, that place of cascading scandals and endless dramas. I thought, when I saw it: *Something's wrong with these people.* They lack a sense of awe, not the awe that leaves you crippled with a false sense of your smallness, but the awe that makes you bigger, that makes you reach higher, as if in tribute to some unseen greatness around you.

That is what the Clintons do not do, perhaps cannot do. They rose to the top but couldn't rise to the occasion. They are not reachers but levelers. They do not inspire.

It's hard to inspire. We live in a cynical age. But it is still possible. Ronald Reagan inspires a number of us, in part because for most of his adult life, when public opinion was running hard in one direction, he was pulling hard in the other, swimming against the tide. He paid the price in many ways, was a figure of fun and derision, was called wild and radical by the leaders even of his own party. But he didn't complain, and tugged and tugged as if with a rope in his teeth until, at the end, the country had come along with him and reached the same safe shore. And when it was over he stood up, smiled, and refused to hate his foes. That refusal—that was heroic, too.

For a number of my friends Jimmy Carter is inspiring, because he is a good man. He left the presidency, went home, wrote his books, started his center, preached his Jesus, built houses for the poor, showed what a fine person can do after the loss of public power, and did it through that most underestimated of didactic tools, the good example.

I thought, seven years ago, that the Clintons might turn out to be inspiring. They had guts, were bright and hard-driving; he was educated, credentialed, a political moderate but not a boring one; she appeared to be something new and interesting, a modern woman who operated with confidence in all the circles of the world.

The stars were aligned for them as they walked into the White House; they had every opportunity to make enduring progress for their country. With the support of the establishments, of big media and Hollywood and more than half of Wall Street; of the big cities, the academe, and a political party hungry for success; with the bell-ringing excitement of the country—a young man, finally, awake and current and bright; with all that and, perhaps most important, a rising economy coming on so strong, in such a great wave that it was beginning to wash the whole country from one end to the other—*with all that behind them, around them, going for them, think of the good they could do.*

He knew the position he was in, and so did she. Because he was a Democrat he could address with courage the entitlement programs; because he was a southerner he could lead, with courage, on race. He could move forward to help the poor in ways his party's left had balked at but that he knew were promising and practical; he could, like Nixon to China, push for technological breakthroughs in the national defense, create a missile defense and a civil defense that would leave the country safer from nuclear, biological, and chemical attacks.

He could speak thoughtfully and like a mature and fully formed man about social issues that divide us; he could give all sides their honest due, help people understand one another, refuse to give in to spite, cheapness, and playing to the base—his base, after all, was going nowhere after twelve years in the presidential wilderness. He could make decisions calmly, wisely, persuasively, treating those who lost with respect and sympathy; he could put the teachers' students before the teachers' unions, reform the schools, re-create the broken system, join the school liberation movement, which would have been hard on his supporters but would have been forgiven in time—because it would have produced progress. In his comportment he could have been modest, tough, and kept the spotlight on the country, not the self. It was the country, after all, that had unleashed the new riches, and that worked out its "social policy" in a billion personal transactions every day. All he had to do was be a sound steward, forgive his foes, give his friends what he could—raise the minimum wage, increase this program, appoint that person—love his country, put it before his passing interests, go the long haul, think long term, be a patriot.

And she—a new kind of woman, exemplar of a rising generation—Hillary could have been a strong and encouraging presence, maybe continuing to work in the world as a lawyer, as Cherie Blair has in Great Britain—a judge, working mother, and "first lady" who is everywhere a figure of respect.

What a presidency this would have been. What a legacy they would have left.

And it wasn't so hard to think, seven years ago, that this is what the Clintons would do, because one assumed on their part good faith, high motives, the courage of the mature, and emotional and ethical stability.

Instead they became what they became. Instead they became what they were, and are. And they produced a presidency regarded by critics across the political spectrum as one long missed opportunity. In the *New York Times,* the historian David McCullough: "William Clinton's presidency was all about him. Real leaders look outward. Clinton wanted to build a bridge to the future. That doesn't mean anything. That's a bridge that runs off into the air. He was never willing to risk his political life for something he believed in." In the *National Journal,* George C. Edwards III, director of the Center for Presidential Studies at Texas A&M University: The Clinton Administration "has not risen to great heights," instead "it has been characterized by incremental achievements and defensive reactions to world events and the initiatives of the Republicans." In the same article, William Bennett on the president: "He degraded virtually everything he touched: the White House, the Oval Office, the staff, the cabinet, the country, the legal process. . . . He is a symbol of decadence."

They missed so many of their opportunities because they acted in a way that was amazingly small time, small bore; they were selfish and cynical and thought small. Their supporters say the headline on their era is "Dow Jones Hits 11,000." But perhaps the real headline on the Clinton era came in the first week of January 2000, and, befittingly, it came in a poll. From the respected pollster John Zogby. Who found that as the millennium began a majority of those questioned said they were ashamed to have Bill Clinton as their president. It was close—42 percent ashamed, 39 percent not—but perhaps even more significant was the fact that when asked to pick the most successful of the past eleven presidents, the respondents gave Bill Clinton the lowest "below average/failure" rating of all but one, Richard Nixon.

What did the Clintons do with their two administrations? They left behind a country more damaged, more removed from its old, rough idealism; a country whose children live in a coarser and more dangerous place; a country whose political life has been distorted and lowered.

This is their legacy. This is the great work of Clintonism.

They have lowered liberalism, undermining a great political tradition by persuading liberals to excuse and accept the corruption of the Clintons as the price of staying in power; thus were left-liberals left defending actions they privately abhor and had once publicly opposed. They have helped tear down heroes from

FDR to Eisenhower to JFK, to defend themselves. "They all do it." But they do not all do it, and they have not all done it.

They have corrupted the Democratic Party by forcing it to be silent in response to their actions, thus lowering all within it from Al Gore to Al From. They have attempted to lower conservatism by smearing it—Republicans and conservatives were not serious people who disagreed with the Clintons for serious reasons but "racists," "Nazis." They have harmed our children by teaching them things through their actions—through scandals followed by public smear campaigns—that adults triumph through dishonesty and divisiveness. They have damaged America's culture by bringing a new level of indecency to our public discourse, and into our living rooms. They have lowered our country's reputation in the world, so that a Swedish socialist, an Iranian fundamentalist, and an Italian communist alike could look at the myriad Clinton scandals and crow about America's amoral leadership. They have compromised the national character by forcing a country to choose between the trauma and dislocation of the removal of a president and seeming to accept that president's low actions. They have lowered it, too, by forcing half the country to have contempt for them, which sours the political system and leaches it of precious measures of idealism and faith. They have damaged American journalism, leaving it full of acrimony, dishonesty, and self-censorship—acrimony between those reporters who want to go after the story and those above them who silence or discourage them; dishonesty from the Clintons' silent promoters in the media, who tilt stories and limit or expand coverage based on whether or not it will help the Clintons; self-censorship from those journalists who see their profession failing in its watchdog role, but who fear damaging their careers by doing, honestly and completely, their jobs.

And in all of this, she has helped him. She was his partner in power; they did it all, together.

Their politics were small time and cynical. They refused to reform welfare until it was forced upon them as a fait accompli by the first Republican Congress in forty years; then they tried to take credit. They refused to balance the budget until it was forced upon them as a fait accompli by the first Republican Congress in forty years; then they tried to take credit. They have failed to attempt to give America protection from nuclear attack and have, at the same time, allowed operatives of the Chinese government and American technology makers to contribute to their campaigns and receive, in return, technology transfers to China that have enhanced a Chinese military structure that may well, one day, be turned on America. (This may become their most lasting and unforgettable legacy, because millions of Americans may someday die because of it.)

They have, finally, left the world able to believe that an American president would bomb another country on the very day one of the victims of his sexual adventurism was appearing for the first time before an American grand jury, in order to divert attention. They have allowed the world to think that an American president would launch a military action against another country, in order to take the country's attention away from the fact that in a few hours he was about to be impeached by the House of Representatives.

There is no greater abuse of power possible to a president—and he chose it.

And she, as always, backed him, and was with him, and supported him.

The Clintons have damaged our country. They have done it together, in unison, and with no apparent care or anxiety about what they have done. And for this reason, for all of these reasons, Clintonism should not be allowed to continue.

And if it is not to continue, the next great battle may prove to be the decisive one, and that is the battle of New York. New York—our empire state, our mini-nation of strivers, geniuses, hard men, tough women, big hearts, winsome creators, first-class strange-o's; of glittering towers and still-shocking unmet need; New York, great home of the old liberal tradition, proud home of Roosevelts, both branches, from TR to ER; New York, proud home of . . . Hillary?

She assumes that liberals and liberalism will carry her to the Senate from our state. But she is not part of their tradition, and should not represent it. This highly credentialed rube, this mere operator, this person who never ponders what is right but only what they'll buy—this person is not towering and generous but squat and grasping. She is, and I never thought I would be able to say this, too corrupt for New York; she is too cynical for the place that gave birth to Tammany Hall.

This race is beyond the issues; she has put it there. If you are a liberal, there are other liberals—vote for them. There are other people who support abortion rights—vote for them. There are other people who care about children, who *truly* care about them—vote for them.

And so the question, for New Yorkers, in this great year 2000, is: Where do you stand?

If all good New Yorkers gather together to resist—all the good liberals and moderates and conservatives—we can of course beat Clintonism back, and end it, decisively.

We can stop it here, in the battle of New York.

You will decide it. Either they will continue to deform and lower our national politics, or they will not. Either Clintonism will continue, or it will not.

It is up to the people of New York.

And that is the great thing about democracy: Before Hillary Clinton gets to decide your future, you get to decide hers.